The Gibbons of Siberut

The Gibbons of Siberut

Tony Whitten

Foreword by Sir Peter Scott

J M Dent & Sons Ltd
London Melbourne Toronto

First published 1982
© 1982, Anthony Whitten

All rights reserved. No part of this publication
may be reproduced, stored in a retrieval system,
or transmitted, in any form or by any means,
electronic, mechanical, photocopying, recording or
otherwise, without the prior permission of
J.M. Dent & Sons Ltd.

Printed and made in Great Britain by
Biddles Ltd, Guildford, for
J.M. Dent & Sons Ltd
Aldine House 33 Welbeck Street London W1M 8LX

This book is set in 11 on 13pt VIP Sabon by
Inforum Ltd, Portsmouth

British Library Cataloguing in Publication Data

Whitten, Anthony
    The gibbons of Siberut.
    1. Gibbons – Behaviour
    I. Title
    599.88′2′09598   QL737.P9
ISBN 0-460-04476-1

Contents

Colour Photographs

Note

A number of local words are used throughout this book and although it is not possible to give a comprehensive guide to pronounciation, some notes might be found useful. The greatest stress is on the penultimate syllable. The letter 'j' is soft as in the French 'je' rather than the English 'joy'. Syllables end or take the first of two consonants following them. An exception is 'ng' which is considered a single consonant. For example, si'ka'te'ngah'*loi*'nak, *bee*'low, si'ka'*ob*'buk, si'*ke*'rei, si'ma'*ko*bu, a'nai'leu'*i*'ta.

Acknowledgments

When a project has dominated over six years of one's life, it is difficult to know whom to thank first for their own particular piece of invaluable advice, help or wisdom. It is, therefore, most reasonable to list the majority alphabetically, thus:

Sjamsiah Achmad (LIPI), Ros and Conrad Aveling, Amsir Bakar, John Birks, Bismark, Paul Burnham, Chuck Darsono, John Fleagle, John Flenley, Paul Gittins, staff of the Grand Hotel in Padang, Janet Barber, Brian Huntley, Jafnir, Kathy and John MacKinnon, Jeff McNeely, Ahmad Mahyuddin (FIPIA), Art Mitchell, Napitupulu, Bill Oddie, John Payne, Jes and Gerard Persoon, Prijono Hardjosentono (PPA), Geoff and Marion Rothwell, Sabri, Saleh Sanusi, Reimar Schefold, Peter Scott, Otto Soemarwoto, Mervyn Smith, Ron Tilson, Anne and Richard Trussell, and Oskar Wessel.

For some, however, special mention is required. David Chivers, my doctorate supervisor at Cambridge University, and his wife Sarah were our most thoughtful, most regular and most helpful correspondents outside of family. Saukani Sarin and his family made our trips to the Sumatran mainland more enjoyable than they could ever otherwise have been, and he kept us sane during the bad times. Alan House provided all manner of help. Mrs Thomas Bata very kindly arranged for the Bata Shoe factory in Jakarta to manufacture special jungle boots for us. The Chairmen of Batchelors Foods Ltd, Cadbury Typhoo Ltd, Crosse and Blackwell Ltd, and Itona Food Products Ltd very generously sent us large parcels of dried foods for use on the World Wildlife Fund surveys. Aman Bulit, Ohn, Abidan and Aman Taklabangan were our long-suffering guides and loyal friends in Saibi. CPPS logging company allowed us to travel on their boat to and from the mainland and offered us great hospitality at their camp on Siberut. Finally, Mollie Whitten, my mother, bore the brunt of my ceaseless requests to send on vital pieces of paper I'd left behind, and suffered most from our months-long periods of being *incommunicado*. My thanks go to all these people.

For one, however, words are not sufficient. Jane was prepared not only to marry me but also to live on Siberut in a camp from which it sometimes took ten days to reach the mainland. Alan House and I had remarked to each other before Jane arrived on the island that if she stayed the course she'd be a wife to be proud of. I am. She took the art of camp cooking with limited ingredients to hitherto undreamt of heights; she kept the complex accounts; she planned and prepared the surveys meticulously; and yet managed to conduct her own scientific study at the same time. This book could not have been written without her unfailing support, patience, tolerance, and repeated improvements of drafts.

Foreword

In the summer of 1970 I visited Basle Zoo in Switzerland. The distinguished Zoo Director Ernst Lang showed me a group of 'dwarf gibbons' or *beelow*, which had recently arrived from a remote Indonesian island called Siberut – one of the Mentawai Islands on the south-west side of Sumatra. 'At lunch,' said Ernst, 'I want you to sit next to a young anthropologist – Reimar Schefold – who has just returned from Siberut, and I think you will find his story interesting.' I did indeed.

Schefold had spent two years with the aboriginal people living on the windward side of the island where landing from a boat was extremely difficult. Those few people from the outside world who had ventured to land there had met with extreme hostility from the 25,000-strong tribe which was isolated from the more accessible leeward side of the island by impenetrable forest. But Schefold had gained their confidence and become a 'blood brother'. And now, he explained, this age-old harmony between the people and their environment was about to be shattered by the timber exploiters who were starting to clear-fell the forest and thus destroy the culture of the people, the habitat of the 'dwarf gibbons' and many other species of animals and plants, a majority of them still unknown to science.

After hearing Reimar Schefold's story, I decided that the World Wildlife Fund and the International Union for Conservation of Nature and Natural Resources should be asked to examine the situation from the point of view of wildlife and environmental conservation, and Survival International from the point of the indigenous people.

This led, among other things, to the studies, adventures and forward planning which Dr Whitten has written about. I first met him when he came as a schoolboy to the Wildfowl Trust at Slimbridge to study the sense of smell in ducks. In 1970 at the age of 17 he was the youngest candidate ever to be accepted for a Winston Churchill Fellowship, with which he travelled round the world for three months looking at wildlife conservation methods.

From 1975 until 1978 he spent thirty months on Siberut island primarily studying the gibbons. Intensive observation of any wild mammal is difficult, but when the subject of your study is a shy primate living over a hundred feet above the ground in dense tropical rainforest, an observer needs a great deal of patience, and determination. It took more than a year of close following for one family of three *beelow* to accept the author as part of the scenery.

But Dr Whitten and his wife were not only trying to learn about the wildlife. They became intensely interested in, and attached to, the people of Siberut. They thought of them as friends and neighbours rather than as elements in an anthropological study and they came to admire their traditionally respectful and controlled exploitation of natural resources in sharp contrast with the reckless exploitation by the logging companies.

In due course the Whittens worked on a plan which had been commissioned for Siberut by the World Wildlife Fund and which would reconcile the seemingly conflicting goals of wildlife conservation and social development. All this and many other conclusions and adventures are described in this book, for which I am honoured and delighted to provide this foreword.

Peter Scott
Slimbridge – 3 June 1981

Dedicated
to the memory of Jack Whitten

Chapter One
Siberut

The lofty, time-worn *uma* was in darkness save for a warm glow shed by the small guttering flames of two makeshift oil-lamps on the veranda. The soft light played on hundreds of off-white monkey skulls hung from the sago-thatch roof further inside, and shadows danced around their staring eye sockets. Gathered around the lamps were Jane and I and twelve members of the Sakuddei, the most traditional clan on Siberut. The predominant sound was from the myriad night-time insects — cicadas, mole-crickets, grasshoppers and beetles — but there was also a low hum of conversation, broken by the occasional expletive and yawn.

'Hey, Aman Beelow,' Aman Uisak, one of the clan's medicine men or *sikerei*, called across to me, 'did you know that men have not always lived on Siberut?'

'Well, yes, I suppose there must have been a time before man arrived on the islands,' I replied.

'No, man did not arrive on Siberut,' he said firmly.

'How did you come to be here, then?' I asked, intrigued.

'It was like this, brother.' Some of the older men and women hawked and spat through the spaces between the floorboards and then they all fell silent and craned their necks to hear what Aman Uisak was going to tell me. 'Long, long ago, when Siberut was the only island in the ocean, there was nothing but forest from Sikapokna in the north to Taileleu in the south. The seas around the island were full of fish and had never been travelled upon by any man. Here in the forests, animals abounded and their life was easy. Life was so easy,' he continued, 'that some animals found they were becoming too crowded. Even the tallest *elagat* trees were crowded and the *beelow* gibbons found there was not enough food for all. So they decided to call a meeting of all the *beelow* on Siberut to discuss the problem. Eventually, after many days of debate they all agreed that half of them should climb down to the ground and try to live there, and the other half should stay in the trees. In time, the *beelow* on the ground changed into men and thus they became our ancestors.'

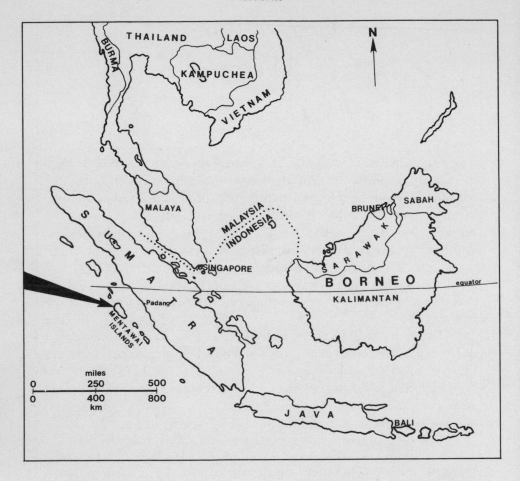

This pretty close approximation to accepted evolutionary theory expressed, as plainly as anything could, the closeness these people feel to the forests that surround them and to the gibbon, or *beelow*, in particular. They are dependent on the forest – they collect food from it, they cut building materials in it, and it provides them with goods that can be sold to the outer, commercial world when they choose to do so. Yet despite these activities their impact is negligible, for the forests have a huge potential to repair small-scale damage, and growing conditions in this region are as near perfect as a plant could ever hope to experience.

I first became aware of the *beelow*, one of the rarest of the nine gibbon species, when its threatened existence came up in conversation during a lunch with Sir Peter Scott and a show-business millionaire. I

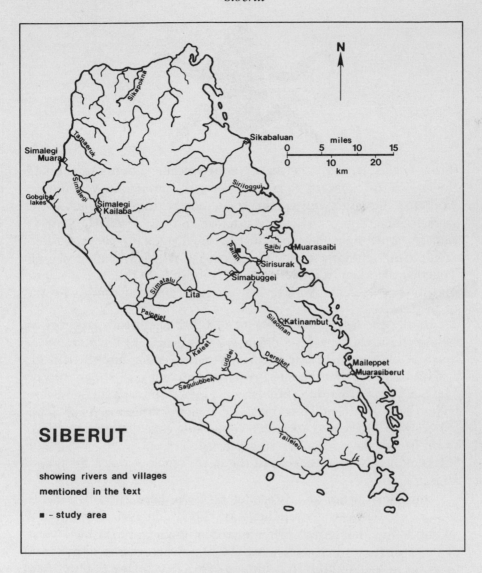

N

Sikabaluan

miles
0   5   10   15
0   10   20
km

Simalegi
Muara

Gobgib
lakes

Simalegi
Kailaba

Sirifoggui

Saibi   Muarasaibi
Sirisurak
Simabuggei

Simatalu
Lita

Painajet

Silaoinan   Katinambut

Kaleat
Kuddei
Dereijket

Maileppet
Muarasiberut

Sagulubbek

Taileleu

SIBERUT

showing rivers and villages
mentioned in the text

■ – study area

was interested enough to attempt some background research but this
didn't lead very far. If authors admitted knowing anything at all about
the *beelow*, the limit of their information was that it lived only on the
four Mentawai Islands off the west coast of Sumatra in Indonesia.

These islands were formed some two hundred million years ago
when a huge geological 'plate' moved slowly but very surely from the
south and collided with the edge of the major Asian 'plate'. Tremendous forces resulted and the plates began to buckle. The major upward

thrust formed what is now the chain of volcanic mountains running the length of Sumatra that end with the infamous volcano of Krakatoa. A large downward thrust created the deep-sea trench to the west of Sumatra, beyond which a minor upthrust formed a chain of islands. In the middle of these, just below the equator, are the Mentawai Islands, the largest and most northerly of which is called Siberut. Together the Mentawai Islands cover about 2,700 square miles, just one third of the area of Wales, with Siberut accounting for three-fifths of the area.

During the last million climatically turbulent years, a series of ice-ages has caused the level of the world's oceans to fluctuate by as much as seven hundred feet, because water was alternately held as ice and then released. During those years, the present-day land masses of Malaya, Java, Borneo and Sumatra were joined and separated, as were Britain and continental Europe. The major South-east Asian land masses were last joined about ten thousand years ago, but the Mentawai Islands have remained separate for very much longer – about half a million years – because of the deep trench between them and Sumatra.

One consequence of this is that whereas there are considerable similarities between the animals of Java, Malaya, Borneo and Sumatra, the animals of the Mentawai Islands are quite distinct. Most of the animals are endemic, being found nowhere else, and many of them are relatively primitive, having evolved at a slower pace than their relatives elsewhere. Among the endemic animals are no less than four species of primate, three monkeys and the *beelow*, and there are no other islands in the world that can equal, or even approach, this level of primate endemism.

From the anthropological literature I could find, it appeared that the people of Siberut represented all stages of a transition between, on the one hand, hunter-gatherers with animistic beliefs and, on the other, developed Indonesian citizens who could blend into a mainland crowd without difficulty. In general, it seemed that the further inland and the further west one travelled, the more traditional was the culture, and the lifestyles of some clans had changed little for centuries. I found the combination of a little-known group of people and rare primates living on a remote forest-covered tropical island irresistible, particularly since they all appeared to be threatened by foreign timber companies.

I resolved to study the *beelow* on Siberut, where logging had had the least effect, and to investigate ways in which the future of Siberut's wildlife might be safeguarded. Even from the early stages, however, it was clear that the status of the wildlife was inextricably related to the island's human inhabitants and their insidious acculturation. The major problems that would have to be resolved, therefore, were how wildlife conservation could operate side by side with the inevitable social and economic development of the people, and how any proposals could be implemented.

The Leverhulme Committee of the Royal Society was the first body to have sufficient faith in my project to present me with a cheque, and this would cover all my expenses for one year. A year was too short to do justice to Siberut, however, so I continued to tell people about the dwarf gibbon of the Mentawai Islands (as the *beelow* was frequently, but erroneously, called in books) and money trickled in.

In 1975 there was a pop record near the top of the charts called 'Funky Gibbon', performed by the Goodies. This added chance of more publicity for the project was too good to miss, and within a fortnight Bill Oddie and Graeme Garden of the Goodies had agreed to appear at a World Wildlife Fund press conference launching 'Project Dwarf Gibbon'. In addition to a generous donation from the proceeds of 'Funky Gibbon' itself, the Goodies' support resulted in cheques appearing out of the blue, which made a considerable difference to the accounts.

Money was becoming more important with the advent of two additional expedition members. Alan House, a soft-spoken fellow student at Southampton University, was to investigate the cycles of fruit and leaf production of the forest trees. Many animal-based projects in the tropics could have gone further had a botanist been present, and I hoped that together we would produce some interesting work. Alan, skinny and tall, had shoulder-length blond hair during his student days. Such a style is somewhat frowned upon by the authorities in South-east Asia, however, and he had to have his hair cut

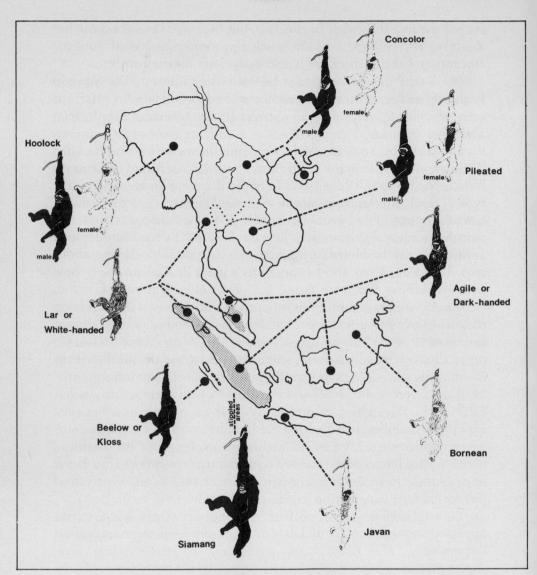

Distribution of gibbons

twice before the result was acceptable. He was to find his height, fair hair, and later his bushy beard, an endless source of fascination and occasionally even terror to the children who saw him on Siberut.

Jane, another blonde, but ebullient and with no respect whatsoever for her prospective gibbon-studying fiancé, was also a Southampton student. She planned to investigate the habits of squirrels that shared the tree-tops with the *beelow*, but because she was taking her finals the year after Alan and I, it was agreed that she would leave for Siberut six months after we had begun to set up the project.

We found that we were not the only ones interested in Siberut. Richard Tenaza, an American zoologist, had visited Siberut briefly to collect basic information about the *beelow*. An American couple, Ron and Janet Tilson, had lived on the island studying some of the primates for about eighteen months up to 1975 and through their and Richard Tenaza's efforts the small Teitei Batti Nature Reserve had been established. The help the Tilsons gave us in planning our budget and on the type of food and the other supplies we would have to buy on Sumatra before leaving for Siberut was invaluable and made our first few months immeasurably easier. Reimar Schefold, a dedicated Swiss anthropologist, had lived on Siberut with the Sakuddei clan for about two years from 1967 and he taught us a great deal about the people amongst whom we were going to live. Robin Hanbury-Tenison, Chairman of Survival International, the London-based organization concerned with the welfare of indigenous peoples, visited Siberut briefly in 1973. As a result of his, Reimar's and Ron Tilson's efforts, a project was eventually formulated that would assist the local government in the socio-economic development of the Siberut inhabitants.

There were a thousand and one jobs to complete before we left England in December 1975, and at least as many more between arriving in Indonesia and taking the perilous, twelve-hour boat journey across to Siberut. Visas, travel documents, letters of introduction, police forms, letters of permission to work, entry permits to the Teitei Batti Nature Reserve, and so forth, had to be collected. Our file of official bumph was tearing at the seams.

We left Padang, capital of West Sumatra, in a rain storm just after dusk accompanied by the head of Siberut's three-man Conservation Department. The twenty-four passengers on the ten-ton boat crouched, sat and lay beneath the heavy, slightly perforated awning above the hold, continually shifting to avoid the new leaks that appeared. In the throbbing bowels of the boat, sacks of rice, sheets of corrugated iron, bags of chilli peppers, tins of biscuits, boxes of nails, bunches of chickens, and rat droppings, jostled against out motley collection of containers, the contents of which represented our anticipated needs for the next few months. When fatigue and the rocking of the boat eventually made us want to lie down and sleep, we found that the arrangements on the grubby boards were rather worse than those suffered by tinned sardines. Mercifully, the restless night passed; cockroaches stopped running up our trouser legs and over our faces, and the sound of retching bodies grew less frequent as we came into the shelter of Siberut. We planned to make the trip between the island and the mainland every three or four months, but at that time the thought of ever repeating it was horrifying.

I raised myself from the planks, slipped under the soaked awning and looked towards our forested destination resting peacefully between the grey-blue sea and the orange western sky. Alan peeped out from under the awning and considered the spectacle with quiet wonder. He sidled past me and wandered nonchalantly towards the stern, attempting, but failing as ever, not to attract attention. The boat's toilet, for which Alan was quite obviously heading, was a hole in the deck surrounded by planks less than two feet high with a bucket of water instead of paper at its side. The young crew graciously averted their gaze while he inexpertly squatted in the box. Even then he couldn't exactly relax because one of the deck hands soon came and sat on the gunwale next to him for a chat.

Siberut was getting ever closer and it was possible to make out leaning coconut trees, gleaming white beaches and waves breaking on the edge of the coral reef. Flying fish erupted from the sea around us, manoeuvring skilfully just above the waves, and a school of dolphins came tantalizingly close, playing and jumping in unison. A guide book could make Muarasiberut, our destination and the island's main administrative centre, sound quite delightful. It is a small village at the mouth of a lazy river, flanked by palm-studded beaches and protected from the tropical surf by a coral reef. Small, colourful boats moor in the harbour and the quay is full of children playing in the sand, and cats, dogs, chickens and ducks foraging for forgotten scraps of fish. Groups of people wearing *sarongs* sit chatting and smoking on their large verandas, and occasionally dugout canoes arrive from the upper reaches of the river laden with exotic fruit, the paddlers wearing flat, broad hats as protection against the sun. Further back lies the rest of the village, watched over by lush, forested hills inland. A large playing field is the scene of a straggling football match with children of all sizes making up the teams. In the early morning as one walks down the coral-paved path to bathe, one sees the schoolchildren lined up for roll call. Further off, someone is bringing in the water buffalo and its calf from a field; this lumbering beast with a small cart provides the only wheeled transport – the gentle pace of village life has no need for anything faster. In fact, Muarasiberut is no more than a latterday colonial outpost, the colonials being mainland Sumatrans and Javanese, and a more wretchedly fly-blown, smelly, dirty and unfriendly place I have never visited. We spent as little time there as possible before entering the island's hinterland.

The first five months on Siberut were a mixed success. We found that we could cope with Indonesian, the *lingua franca*, quite well, but this is no particular credit to us since it is supposed to be one of the easiest spoken-languages in the world. It was created just after Independence, predominantly from a central Sumatran language that was frequently used in trading, to facilitate communication and to help unite the vast republic. There are only three tenses; no genders, declensions or conjugations; and even before leaving England we had mastered such useful sentences as 'Is that a fat man? No, it is a big fish'. Many of the people on Siberut speak little or no Indonesian, however, and so learning the local language was a priority. This proved rather

Young *beelow*

more difficult, particularly when a set of mischievous children decided it would be great fun to teach us nonsense words and have us think that their peals of laughter were caused by imperfect nuances of pronunciation rather than by the gobbledygook we were actually talking.

With the assistance of the local Conservation Department we located a suitable study area within the Nature Reserve, about two hours by dugout from the village of Sirisurak in the Saibi river basin. Our most important task was to open up the forest by cutting paths and mapping them using a compass and a knotted rope. In the meantime, however, between us we caught malaria, had flaming fat poured over a foot, had the first of innumerable fungal infections, and began to learn the hard way about the pitfalls of a forest life.

We also caught sight of some *beelow*.

Chapter Two
Early Days

Over the Paitan river in front of our house, a male *beelow* started to sing. It was four in the morning and I hurriedly packed my shoulder bag, tripped over Jane's boots and walked across the clearing to the forest. As a town-dweller for most of my years, I had rarely had the opportunity of looking at a truly star-filled sky and so took great pleasure in standing and staring for a little while. Even with no moon, few parts of the sky were actually black, being rather a dark grey lightened by a multitude of shimmering stars, and since we were on the equator, I could see a mixture of familiar and unfamiliar constellations. Orion hovered above the river and the Southern Cross hung just above the coconut trees at one end of the clearing.

There had been heavy rain each morning for the past week and the clothes clinging to me were heavy and damp. The temperature was about 70° Fahrenheit, the same as a warm summer's day in England but here, with one hundred per cent humidity, it felt bitterly cold and I shivered, feeling deeply envious of Jane still sleeping in a warm bed.

The distant *beelow* stopped singing after a few minutes, possibly because it was too cold for him or because none of his neighbours had joined in. Just within the forest the path was a quagmire and soon my feet were soaking, muddy water squelching through the canvas of my boots as I walked. The path was bad for several reasons: it was used by us several times each day; it had been very wet of late with up to three inches of rain falling in an afternoon; and it had been adopted by the dozens of feral pigs that lived in and around our campsite. They tended to follow most of our paths around camp (or else we had followed theirs when we cut them) and cutting alternative routes was only a temporary solution.

It was only a short distance to the foot of the major spur of Teitei Bulak, the hill behind our house. Soon after I began to climb, wishing dearly to be back in bed, the path looked different; someone had felled an *ariribuk* palm across it, presumably during the previous afternoon when I had returned from a disappointing day in the forest by a

11

different route. *Ariribuk* was one of our biggest hates in the forest for, as its scientific name, *Oncosperma horridum*, suggests, it is singularly unpleasant. The eight-inch-thick trunk is covered with black, very sharp, two-inch spines all the way from ground level up to the leaves that fan out about sixty feet above the ground; even the stem and main rib of the leaves are covered in spines. Odious though it is, it does find some uses because the exceptionally hard wood on the outside of the trunk makes extremely good firewood and, once scraped clean of spines, makes long-lasting floorboards. At the heart of the leaf bases lies the 'cabbage' or primordial leaf shoots; about twelve by one and a half inches of a delicious vegetable reminiscent of white cabbage, that can be eaten raw or boiled. It seemed that the trunk in front of me had been felled for the sake of its 'cabbage' but I wondered how on earth this had been extracted since the people on our river wore no shoes to protect them from the spines.

At the top of the spur I stood still for a few minutes and listened carefully for the plaintive songs of male *beelow*. There were none to be heard, so I turned left down the spur towards the Sibosua river. The torch provided only a narrow tunnel of light in which to walk and I had to rely on my knowledge of the path to avoid stumbling. I held the torch at head height in front of me to avoid getting my face covered in spiders' webs and spiders. Some of the webs were enormous, often four feet across, yet they were built overnight and could last for days. Others were so tough that walking was checked. I'm not a great fan of spiders at the best of times and trying to remove a horned spider with a body over an inch across from my sticky, sweaty face in the dark was not my idea of fun. So I kept the torch held high.

I was aiming to reach a little clearing where Aman Mynoan, a *sikerei* from the Sagaragara clan, our neighbours on the Paitan, had recently felled a *katuka* tree for a dugout canoe. This meant wading through a couple of usually shallow, meandering rivers which by virtue of the recent rain were now over knee high. While crossing them, the torch spotlighted little silvery fish and dark crayfish with red reflective eyes darting away as they sensed my footsteps. In order to avoid these daily soakings, we had previously cut small trees to form bridges over the streams. But the streams flooded badly at least once each month, and the bridges were repeatedly washed away. I reached the little clearing by 5.30 am and since none of the *beelow* were singing, started on my breakfast of sweetened black rice, squatting on what remained of the *katuka* tree.

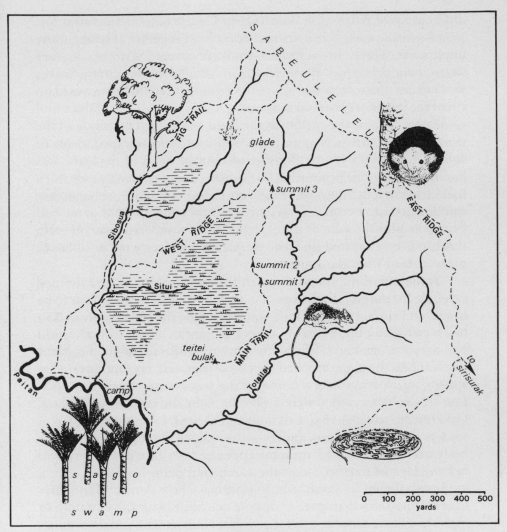

Study area

The sky was beginning to lighten and I could see small bats flitting in and out of the shadowy clearing like satanic butterflies, occasionally dropping close to the ground by some rotting hollow logs looking for roosts. The birds around me were waking up and starting to call. Hill mynahs (the mynah birds of pet shops) were chorusing coarsely, employing a wide range of unlikely sounds including one that was so infuriatingly similar to a young *beelow* throwing a tantrum that one might be forgiven for thinking it was a direct imitation. The black and iridescent-blue Asian fairy bluebirds and the black, fork-tailed drongos also began their choruses and by now most of the animals in the forest were awake. The diurnal ones were preparing to live another day and the nocturnal ones preparing to sleep; the intermittent scufflings behind me were probably caused by forest rats returning to their burrows after a night of foraging.

Dawn was marked, as always, by the deafening chorus of a particular type of cicada. Just as dawn broke, one of these long-winged, dumpy insects started to make a buzzing noise that creaked and groaned as though it were rusty. Eventually, after several abortive tries, it completed the first, long rising note and then gave shorter rising notes for several minutes while its neighbours joined in. Suddenly one of them stopped and I became conscious of a small ghostly shape passing through my field of vision only to become invisible again as it landed on another tree trunk to resume calling.

During this ear-splitting din I left the clearing and started walking slowly northwards. I had found a *beelow* group sleeping near here several times before, and on the few mornings that I had located them in the actual tree they used for sleeping, they had left it around dawn. This was presumably the time at which they could see well enough not to misjudge distance and fall. Also, for the first few minutes after leaving the sleeping tree, the *beelow* seemed to travel to a fruit tree, and I hoped to be able to detect, if not the *beelow* themselves, then at least the movements of the trees in which they were travelling.

I walked slowly along the flat path looking left and right into the tree tops. The forest here was totally different from anything I had seen in Britain or other temperate regions. The overriding impressions were of the vast quantity of plant life and of the massively vibrant presence that it seemed to emanate. A woodland in Britain might contain as many as ten tree species – here there were easily ten times that number. In Europe few trees reach over a hundred feet, but here the majority exceed that and some were over two hundred. For some of those giants, fifteen men standing around the bases of the buttresses would have had difficulty reaching out to touch fingers.

The trees were obviously the major but by no means the only element of the forest. In the confusion of the complex upper canopy most of the leaves looked very similar – oval, about six inches long with a drooping tip. On closer inspection, however, it could be seen that some belonged to opportunistic epiphytes, plants such as flamboyant orchids which use tree branches as a place to grow, and some to parasites such as mistletoes which actually penetrate the tree and derive some nourishment from it. Other leaves proved to belong to lianas that grow up from the ground, attach themselves to the tree by hooks, suckers and knots and climb to the top of the forest canopy, leaving a long dangling stem of the type Tarzan was wont to use. Yet others were stranglers, plants that grow from the top of a tree and eventually cover the trunk of the host with their roots to form a living coffin. In time the host dies and decays leaving a hollow 'trunk' of fused roots. Some leaves in the canopy were conspicuously different; large dinner plates, soup ladles, Adam's protective clothing and baseball bats were all imitated and the graceful, ever-moving feathery fronds of the climbing palms or rattans were ubiquitous. Rattans make their way to the top of the forest canopy using long, viciously hooked whips that wave about in the wind until they become attached to some support. The hooks bend backwards so that as soon as contact is made, they dig deeper into the support. Even after only a few months of walking in the forest, our shoulders and hands bore witness to the stinging efficiency of these rattan whips which often seemed invisible against all the other elements of the dense vegetation that too frequently barred our way.

In the darkness of the lower canopy where I was walking, there was also a considerable amount of plant life. Here were the saplings of the next generation of forest giants and all the aerial roots and stems of lianas, stranglers and rattans. Here too were mature specimens of palms, short trees, shrubs and a few herbs that have become adapted to the low light levels. Some trees near me were flowering; small, undistinguished yellow flowers that, judging by their rather pungent smell, were probably pollinated by flies. The predominant impression here was of vertical lines. The few horizontal lines were found only in the form of fallen trees — those freshly fallen still covered with epiphytes and moss, all slowly dying in the comparative darkness, and those from years ago which now, by the action of termites, were scarcely more than longitudinal mounds of russet earth.

But the sensory nightmare of living profusion wasn't just visual. At no time of day could the forest have been said to be quiet. Everything seemed to make some noise — leaves brushing against each other in the sultry breeze; ants inside their *maché*-covered leaves hitting their heads up and down to warn you of the bite you'd get if you tampered with them; frogs making all manner of booms, gulps, buzzes and chimes at volumes quite out of proportion to their size; monitor lizards uttering penetrating barks, invisible grasshoppers, crickets, cicadas, beetles and bees whirring and humming; birds with foghorns and birds with discreet tweets; primates with dramatic whistles and yelps and thousands of others we never identified. Some of the forest smells were clearly identifiable, such as the fruity aroma of fresh primate dung. But others – fragrances of steaming porridge, smoky bacon crisps, fresh-cut apple, carrots and German sausage that pervaded the air around certain trees – reminded me cruelly of foods I wouldn't be able to eat for another two years.

I traipsed on, glimpsing movements in the trees which, on further investigation, I always found to be caused by birds. I tried to walk as quietly as I could, but the large leaves from the abundant dipterocarp trees were so crisp that it was impossible not to sound like someone walking across a huge bowl of cornflakes. Suddenly I felt a stab of pain – like a cigarette burn – in my crotch. Dropping everything, I fumbled urgently with my shorts and managed to extricate a wriggling, orange-brown and yellow leech about two inches long. These blood-sucking relatives of the earthworm were very common in the low-lying areas of the forests and they were especially active when the forest floor was wet after rain. They hung beneath leaves next to paths and as they felt the vibrations of a footfall, so they waved around excitedly on the offchance of making contact with a leg. Once on, they looped towards somewhere warm such as an armpit, crotch or the inside of a boot, and bit. The Siberut leech is one of the very few species in South-east Asia that hurt when starting to push their Y-shaped tongue through your skin. This proved a very obvious warning of imminent bloodsucking and it was generally possible to pull the leech off before it started, particularly because our leg hairs became sensitive to their looping progress. Other types of leech have the infinitely more sensible scheme of injecting not only an anticoagulant, but also a mild anaesthetic so that the bite isn't felt. On Siberut we rarely provided meals for leeches except on days when our legs and feet were scratched by rattans, when the additional sting of a leech bite went unnoticed. Only then did we discover either an engorged, taut and only vaguely animate capsule hanging from us, or an obstinate flow of blood penetrating our clothes that might last for hours.

When I began work, I knew that Ron Tilson had reported that *beelow* groups in his study area, five miles south of ours, lived within home ranges of less than thirty acres. A *home range* may be thought of as the area in which an animal travels during its daily activities; a

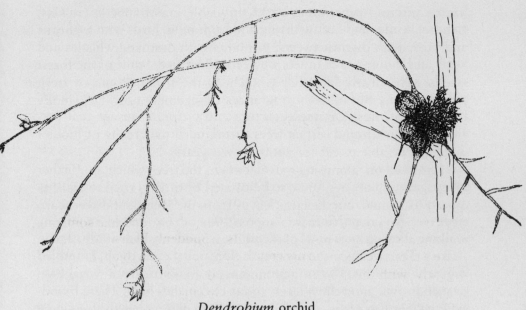

Dendrobium orchid

territory is that part of the home range over which it has exclusive use and which it will defend against intruders. I was influenced by his findings and to start with was able to draw lines on our map of the study area around my few sightings of *beelow* to form home ranges of about thirty acres. It seemed so easy. Soon after that, however, some of my *beelow* crossed the 'boundaries' and so I had to redraw the lines. I had to redraw the map several times and then realized that the two study areas must have been fundamentally different and that the home ranges of my gibbons must be closer to the size reported for other species. The resulting picture was such a mess that I threw it away and waited until I could habituate a group.

To study gibbons well, it is essential to habituate the animals so that they act normally while being observed and so that you can move around beneath them to find a good vantage point. If they are not habituated your views tend to fall into two categories: either the animals are virtually invisible because you are having to hide to avoid them seeing you; or else you see black oblong shapes surrounded by flailing limbs as they race away from you. Habituation is a lengthy process and most of the time, as was happening this morning, was spent wandering around the forest looking for the animals. When they were eventually located, sightings tended to be embarrassingly brief.

Several groups from three other gibbon species had been habituated previous to my study and all those had been in Malaya. I was trying to habituate the first group of *beelow*. Only in this way would I be able to watch their family life and accompany them through the forest.

The sky had clouded over and the air felt quite chilly. There were still no signs of *beelow* around me so I sat down with my notebook on a convenient log and wrote notes on the morning so far, in the hope that even these negative observations might one day prove useful. I could see about twenty-five yards looking horizontally; about twice this distance looking obliquely upwards. The chances of a *beelow* group being in my field of view at any one time were pretty slim, but the chances increased if I just sat still for a while. By sitting still, however, it was possible to discover one of the disadvantages of wearing short-sleeved shirts and short trousers, and that was mosquitoes. The sweaty heat of my body clearly attracted them and while they whined like miniature jet-fighters in a grey dancing cloud surrounding me it was possible to distinguish two sorts. One had a black-and-white-striped, hump-backed body and was described glibly in the health guides as a 'nuisance mosquito'; the other was a larger, rather sleeker and sly-looking creature whose body was as straight as a guardsman's, and was probably carrying a malarial parasite in its mouth parts. There was no discrimination made between them, however, when it came to swatting. Insect repellent was excellent in camp but soon washed off in the forest where we were continually brushing against damp vegetation.

I hadn't realized that I was craning to hear any sound that might come from an animal until I felt my ears pulled back and became conscious of a discrete but heavy rustling in the leaves just down the slope behind me. It wasn't a gibbon, that was certain, neither was it sufficiently brutish to be a feral pig. To get a better view I walked up the hill a few yards and crouched behind a dense clump of *ariribuk* from where I could peer between the fronds and see without being seen. The source of the noise was now quite close and as I looked down I saw, only about two yards away, a pangolin or scaly anteater. I couldn't have been more surprised or delighted. Pangolins had been collected on the islands some fifty years before but I never expected to see one of these relatively uncommon, usually nocturnal, beasts.

As I watched, it sniffed around fallen *ariribuk* trunks, forcing its shielded snout under pieces of wood to move them. The whole of the upper half of the body and all of its tail was covered in overlapping horny scales giving it a rather prehistoric, reptilian appearance. In fact, the scales are not bony, but agglutinated hairs, rather like the horn of a rhinoceros. It had hefty claws on its feet and was using these to tear apart fallen wood. Eventually it found an ants' nest and it began flicking its sticky tongue in and out of its toothless mouth picking up the small scurrying ants. Someone once counted about 200,000 ant-workers and pupae in a single pangolin stomach. If that represented a day's diet (though probably an underestimate) a pangolin would have eaten more than 100 million ants in eighteen months. The pangolin's amazing tongue is one and a half times the length of the head and body and controlled by muscles anchored to its hips.

The pangolin began to move away, snuffling and scratching as it went, and knowing how much Jane and Alan would want to see it, I dashed out from behind the *ariribuk* and was astounded by how fast it could move, looking rather like a mechanical otter. Before I could reach it, it began to climb a tree and I only just managed to get hold of its scaly tail. It had climbed the tree like a caterpillar; first its front feet and then its back using its tail as a support. I needed both hands to pull it down and as it reluctantly descended, the bark of the tree was furrowed by its claws. With one final tug it was free of the tree and I held its rolled-up body aloft by the tail like a shopping basket. The pangolin is the only mammal in South-east Asia that possesses a truly prehensile tail and employs it to hang from branches when investigating ants' nests in trees.

I was now able to examine it at close quarters. Its belly, inside leg and the sides of its face were covered with soft, short light-brown hair and its small rheumy eyes looked anything but beady. Its nostrils, which were right at the end of its snout, twitched excitedly and suggested that its sense of smell was probably rather more acute than its eyesight. Its ears were no more than small ridges. As I stroked its face, its nostrils closed up, a membrane shot across its eye and the ear canal constricted; reactions that were all probably very useful when ants were crawling angrily over its head.

I had heard of a hunter who once clubbed a pangolin he had snared, slung it across his neck, and taken it home to cook. He never arrived; the next day he was found on a jungle path, strangled to death with the pangolin sleeping curled around his neck. The pangolin had probably merely been stunned, and on coming round had been frightened and curled up like a hedgehog. Once in this position a peculiar muscle lock is engaged which can be controlled only by the pangolin.

Bearing this in mind I continued to carry my trophy by its tail.

Back at camp, Jane had just returned from a fruitless morning looking for squirrels and Alan was analysing some tree data with Stumpy, the camp cat, curled cutely in his lap. Stumpy had been acquired in Muarasiberut to scare, catch or taunt the increasingly lively rat population in our house roofs. He walked on ridiculously short legs as though he were stepping into a series of holes, and had a tail that looked as if it had been caught in a door and the remains tied into a knot, but Alan had been assured he was born like that. They all came to admire my prize but Stumpy gave it one cautious sniff and then tore into the safety of the forest.

'We'd better keep our voices down,' Alan advised, 'I heard some women over the river a little while ago and the last thing we should do is let them know what's here. Pangolin is supposed to be a real delicacy and Aman Gogolay said it tasted like chicken only much better.'

We couldn't hear any voices but we could make out the sound of a dugout being poled up the river towards us. We took some more pictures of what was now a tight, horny ball and then returned it to the forest. I resumed my gibbon search.

I walked up the hill as far as we had explored and then slowly down the Main Trail. Strangely, the higher in the study area you climbed, the steeper became the slopes and the thinner the soil, but the trees grew taller. We measured one huge specimen with a telephoto lens and found that it would have dwarfed Nelson's Column; it was almost two hundred and fifty feet tall and its boughs had a spread of about eighty feet. The sky was still overcast and a light breeze was starting up, but on the forest floor scarcely anything moved. Along much of the Main Trail there grew a plant similar in appearance and habit to dog's mercury in Britain. In some areas it gave the appearance of the carpet in an oak wood's bower, but it was necessary to raise your eyes only a few degrees for the resemblance to be ended abruptly. In some areas there were some circular patches where these short plants were brown and dead; I made a mental note to ask someone about this.

Dotted around the study area we had built little shelters made of a frame of small saplings resting against a thicket of *ariribuk* and then covered with palm leaves. They weren't totally waterproof but they provided some shelter from heavy rain and weren't as sweaty as plastic waterproofs. I had been meaning to build a shelter here at Summit Three for some time so I cleared away the black *ariribuk* spines from below a nearby thicket and looked around for some saplings of the right size. I found one about twelve feet high and cut into its base with a diagonal slice of my *parang*. As it toppled over, I heard a soft hooing sound above me. *Beelow*!

I darted beneath a small tree with a dense crown and tried to see them. There, about a hundred and twenty feet above me was a pouting *beelow* looking affronted, its heavily-lined black face contorted into a deep frown; the first sighting I had had for a week. Not only was all of its exposed skin black, but all its silky fur was too. Gibbons are the smallest of the apes, and this one must have weighed about twelve pounds, as much as a tom-cat, and where one might have expected a tail there was nothing. I could see only this one and it was obviously looking straight at me. Its rounded body was hunched over a thick bough with its chest as low as its feet and its long skinny arms held out as though it were about to launch into a death-defying swallow dive. Feeling pleased that it was staying in one position and not merely rushing off as most *beelow* did, I tried to look nonchalant and uninterested. I pulled some leaves off the tree next to me and sorted through them, pretending to look for insects. Occasionally I put a leaf to my mouth to pluck an imaginary insect from it. I scratched myself and looked up as often as I dared to make casual eye contact, but it was the eye contact the *beelow* seemed to dislike, and after four minutes it could stand no more. Uttering a few slightly louder, faster hoos, it leapt off the bough sideways and appeared to be suspended in mid-air for a second before it made a spectacular landing in a neighbouring smaller tree. In front of this *beelow* I saw two others tearing off southwards, one smaller than the other, suggesting that the first to see me had been the male.

There was no point in pretending I was disinterested any longer, because the most important task was to keep with them as long as possible. In the hope that I would still be close to them when they had to stop and catch their breath, I ran along the ridge path which I had tried to clear well enough for just such an eventuality. The *beelow* were on my right, travelling parallel to me and, since the trees in which they were moving were rooted some way down the slope, we were nearly level with each other, although some eighty feet apart. I managed to get between the male and the female who was now carrying the young one, so that the male had to force himself to advance towards me slightly. I just stood and watched him as he did so, trying desperately to appear unexcited. How he managed to keep one eye on me and still leap, climb thin vines and bounce off tree trunks I shall never know. I applauded his nerve. I hurried on again, trying to pass him, for we were approaching Summit Two and the *beelow* were veering to the right even more and starting to follow a subsidiary ridge along which I had yet to cut a path. All three animals were hooing, and now and again the male sang loud one-second notes at an even pitch. The female slowed down and sang the same sirening notes, but when I moved a little sideways to get a better view of her she raced off, and the male sang a peculiarly eerie descending trill. He now moved closer to me looking as worried as ever and I called softly to him, 'Sam, Sam, shut up for heaven's sake. I'm not going to hurt you.'

Exactly why I called him Sam I don't know, but the name stuck. He heeded me for a few seconds but then uttered another shrill trill as he swung the closest he had ever been to me and then climbed up to a distance he obviously regarded as safer. He sat down with one arm by his side and the other holding on to a branch above him, but still pouted and hooed at me. He kept his right hand where it was and slipped off the branch so that he was now hanging by one slender hand with nothing save air between him and the ground over thirty yards below. He gave some more sirening notes and followed them with another trill. He had scarcely finished when the female trilled too. These loud calls were probably a form of mobbing, such as you can watch in birds, in which the animals try to say, 'Look, we've seen you now. We're watching your every move so don't think you stand any chance of harming us.'

The female sounded some way away now and since both Sam and I feared we might get left behind, we rushed down the ridge. With no path to follow, I was much slower than he was and had to cut a rough trail as I went. In some places this was scarcely necessary but in others, where a tree had fallen and brought dozens of rattans with it, I was scratched and torn on the flailing hooked whips. I recalled the day Alan had caught his eyelid on a rattan whip while he was running home in the rain so I moved cautiously but steadily, and fell further behind the *beelow* which were now heading across the swamp. After only a minute I was thigh-deep in wet peat and felt the chase would in fact be better served if I took a compass bearing on the *beelow* and guessed where they would emerge on the other side of the swamp. I reckoned they would appear at the bottom of the West Ridge so I ran back up to Summit Two and around the other three sides of the square. I would have made faster time had I not slipped and reached out my hand to grasp a small trunk that turned out to be an *ariribuk* stem. They always seemed to be strategically placed by patches of unstable mud. The spines went into my fingers and the base of my thumb and although I was able to get some of them out, others broke off at skin level. I began to feel pins and needles all over my hand; I would have to learn to fall more carefully.

I reached the bottom of the West Ridge in twenty minutes – the same length of time I had spent with the *beelow*. I was quite pleased with the effort since it was longer than most of my other sightings to date and I had added new information on what parts of the forest they used. Nothing moved around me so I sat down, rested my back against a small tree and opened my food box for a late lunch – a cold offering of last night's supper – mackerel and fried rice with chilli pepper. I ate it with my fingers, not really because it was the custom in West Sumatra but because utensils had a habit of falling out of my bag when I was chasing after *beelow*. As I ate, a few grains of rice fell off my fingers onto the ground. Within a minute, a large black solitary ant found them, hoisted one above its head and hurried off. Shortly afterwards another ant arrived and in three-quarters of an hour no one would have known I was a messy eater.

The tree tops were all beginning to move as the wind started to freshen. This almost always meant rain but I stayed where I was for the time being. Soon, however, I heard a sound like a huge crowd clapping and the noise became louder and louder, closer and closer until I could make out the warning sound of individual raindrops on leaves close by. It was only 3.30 but the forest was dark and gloomy, and getting colder by the minute. Then suddenly the clouds discharged their thunderous load, the tree leaves bowed under the increasing weight until they could shelter me no longer, and I ducked beneath the palm-leaf shelter I had built some days earlier. I'm not sure why, but the deafening sound of torrential rain in the forest always made me sing. Bach and Handel arias were followed by negro spirituals and interspersed by pieces I could never remember the name of. After twenty minutes or so the rain had found a way through the palm leaves and I had to manoeuvre my body to miss the ever-increasing number of cold drops. I felt like a magician's lissom assistant must feel when she has to contort her body inside a tall box as the magician slides swords through it. Eventually, I couldn't hold the necessary position any longer and so resigned myself to getting soaked. The rain was as heavy as ever, showing no signs of relenting, and innumerable squadrons of determined mosquitoes had found me. In such conditions and for some time afterwards there was no hope of finding any *beelow*, so I put my notebook in a plastic bag and started off home.

It was still raining hard when I reached camp, two houses nestling in a clearing below the study area's forested hills. 'House' is, perhaps, a misnomer because each was little more than a thatch-covered room, on three-foot stilts to keep it clear of the flood waters. The design was an adaptation of the local field houses or *sapos* to allow for our greater height; after living in a *sapo* for several weeks when we first arrived Alan and I felt we might never be able to stand up really straight again. Between the houses was a small garden where Jane was trying hard to grow vegetables despite the ravages of grasshoppers, deer and marauding pigs that somehow managed to penetrate the stout palisade we'd built around it. At the side of the house that Jane and I occupied, the bamboo guttering was working well and had already filled the plastic dustbin on the veranda, so the flow of lukewarm water was being wasted. I tossed my bag onto the veranda, stripped off and had a wonderfully invigorating shower.

'Well,' I said rather smugly to Jane, who was sewing up her slowly decaying pair of shorts again, 'what did *you* manage to see today?'

'More than you think from the way you're talking! I watched a *jirit* squirrel for ninety minutes.'

'You did what?' I exclaimed.

'I watched a *jirit* for ninety minutes,' she repeated. Now it was her turn to look pleased with herself. After two months of patient watching in the forest, her previous record for a continuous observation of this gerbil-sized, light-brown squirrel was just two minutes – if you exclude the dead one Stumpy had proudly brought into the house two weeks previously.

'You know how we've seen *jirit* eating bark on a tree for a while before they skip off and we lose them in the undergrowth,' she continued, 'well, that didn't happen today.'

'What did it do instead, then?' I asked.

'Nothing.'

'What do you mean "nothing"? It must have done something, surely.'

'No,' she asserted emphatically. 'It just stood on a low log up on the Sabeuleleu ridge. I started off sampling its behaviour at one-minute intervals, and then every five minutes, but it just did nothing.'

'I wonder what it was doing nothing for,' I mused. Jane looked at

Jirit

me in the scathing way she has when someone tries to read too much into animal behaviour.

'Maybe it had nothing better to do.'

'Are you sure it wasn't just scared rigid of you?' I asked.

'Yes,' she said firmly, 'they crouch down and look furtive when they're scared. This one just stood there in one position for ninety minutes and then sauntered off. Tell you what I did see moving though,' she continued enthusiastically; 'while I was watching the *jirit*, I caught a glimpse of one of those mottled brown frogs jumping frantically across the path. Then I saw a dark brown snake chasing it. After that all I could see was the frog leaping through the under-growth.'

So much for my twenty minutes watching *beelow*.

The three of us had been invited to visit Sirisurak at Christmas to join in the festivities. So, on Christmas Eve afternoon we loaded our unkeeled, eighteen-foot dugout with presents and various belongings crammed into strong plastic bags and began paddling downstream. The river was very high after a couple of stormy days and we antici-pated an easy and extremely rapid ride. At that stage none of us had quite got the hang of steering and by over-reacting and then over-compensating we bounced hopelessly between the banks and negoti-ated the bends in a number of hair-raising and unconventional ways. The river was almost bursting its steep banks, and branches that were normally some yards above our heads were now brushing our hair and dropping large spiders onto our stiffening legs. In the eddy currents of one bend we saw a long, grey, bloated and horribly fetid mass that must once have been a mighty python.

After about an hour we rounded yet another bend to see a thick tree trunk held fast between the rust-coloured muddy banks, and a vibrating dam of twigs, branches and boughs in front of it. We headed, more by luck than by judgement, towards the right bank where we reckoned it would be easiest to haul the dugout over and as I was in the front, I climbed cautiously onto the trunk. Jane and Alan fought hard to keep the hull parallel to the current, but their battle had scarcely begun before it was lost and the back of the dugout swung round to hit the trunk. A frothy wave rose at Alan's side and barely a second later the dugout had been sucked down, and the plastic bags escaped, bobbing away downstream. I dived off the trunk after the bags and managed to retrieve and wedge all but one of them between roots jutting out of the sheer bank. By the time I had pulled myself back along the bank to the offending trunk, Jane and Alan had twisted the uncooperative dugout round so that water flowed around rather than into it, and we managed to manhandle it over the trunk and into clear water; a task made no easier by the bank being covered in calf-high *alalatek*, a murderously superior relative of the stinging nettle.

We retrieved my caches and paddled very carefully on to Sirisurak, finding the stray parcel trapped in a twig-filled eddy. In our muddled efforts to park our wretched craft at the village we nearly swept on to the coast, causing great amusement among the line of wide-eyed children watching from the grassy bank, some of whom shouted incom-prehensible instructions on the finer points of dugout manoeuvring.

The village of Sirisurak was established in about 1950 by German Protestant missionaries and sanctioned by the Government as an attempt to make the surrounding inhabitants forsake their traditional, scattered, clan-centred way of life. The original church was demolished shortly after we set up our camp and the new white-painted church, thatched, raised on stilts, and with a broad veranda like all the other buildings, was the focal point of the village. Around it lay three parallel 'streets' comprising about thirty houses, roughly one for each nuclear family, but the majority were used more as weekend cottages than as permanent houses. Almost no one connected with Sirisurak lived further than three hours' dugout ride away and the church had the dual role of centre for worship and weekly meeting place. Most of the services were taken by the villagers themselves, but now and then the minister, born in the southern Mentawai Islands, paid a pastoral visit.

We went directly to the house owned by Aman Bulit. This thickset, genial man had worked as assistant to Reimar Schefold when he lived with the Sakuddei, and had helped us greatly in setting up camp and adjusting to a totally new lifestyle. He had a peculiar understanding of the Western mentality, without shedding any of his indigenous pride, and with his fluent Indonesian was invaluable at answering our never-ending stream of questions and at scolding us gently whenever we unknowingly did or said something rude or tactless. He and his wife, Bai Bulit, were childless and the village had decided not long before that the couple should adopt their second cousin, a young girl called Bulit. The prefixes 'Aman' and 'Bai' mean 'father of' and 'mother of' respectively and are followed by the name of their first-born child. So, Aman Bulit was being saved the embarrassment of not having a name befitting his age.

He was startled to see us looking so wet and bedraggled and he called to Bai Bulit to make drinks and eats. It soon seemed as though half the village were in or around the house and a fair proportion of them stared fixedly at Alan or told their astonished friends how they had seen him actually having to stoop to get onto the veranda. We told the assembled crowd what had happened and they gave us barrow-loads of good advice for our dreaded return journey the next day.

We were all summoned to church after supper at ten o'clock. Most of the men were dressed in their smartest Western-style clothes, but tattoos and beads showed at their open necks. The women were dressed in the style of West Sumatran ladies: colourful batik *sarongs* with lacy, long-sleeved blouses, and chiffon scarves over their shoulders. The church, decorated with frayed palm leaves and even a 'Christmas tree', was absolutely packed with people, dogs cringing from kick to kick, a few stray cats and a couple of confused hens.

To begin with children were led one by one, like lambs to the slaughter, to the raised stage where they had to recite a verse of scripture they'd learnt. Many suffered from stage-fright before completing the first line and were subjected to several excruciating minutes

27

of embarrassed squirming before an elder gave them a prompt. Next Aman Bulit conducted his small choir in a number of carols, one of which was a translation of 'Silent Night', prayers were said and hymns were sung. Each line of a hymn was called out by an elder before being sung very loudly and with great gusto by the congregation. When the declamatory sermon began the verger locked the church door, pocketed the key, and then patrolled the aisle poking anyone who seemed, despite the volume of the delivery, to be dropping off to sleep. We could only guess at its contents, since in common with the rest of the service it was in Mentawaian and the vocabulary was totally beyond us (there being no mention of pigs or bananas). Then, after more hymns, the collection was taken. This caused the most unholy commotion as people jokingly begged loans off their friends, the elders and the verger, and the bag was dropped several times before it completed its hazardous journey around the congregation.

The next morning we joined another, less frenetic, service after which we waited while a dozen pigs were slaughtered and their meat and other edible parts shared out with painstaking fairness into sections of banana stem, one for each household. We had a delicious Christmas lunch of pork in a richly spiced sauce, rice and sago with Aman Bulit, his dotty old aunt Lasui, and Bai Bulit. Christmases we had grown used to were family occasions but in the absence of genuine family we couldn't have wished for a friendlier or more welcoming time than we had in that house. We were slowly becoming accepted as an integral part of the landscape and for that we were extremely grateful.

Refreshed by the Christmas break, it was back to a crepuscular routine, and on Boxing Day I walked directly to the West Ridge, being scared rigid by the unexpected noctural wanderings of a herd of feral pigs on the way. The clouds were dispersing and sunlight was beginning to stream into the forest for the first time in days. As I continued north-east so the light caught the drips of rain held at the ends of leaf tips, making them sparkle like jewels. Spiders' webs were like plaited rows of pearls with their owners like grotesque pendants. The whole forest had a hazy, almost mystical appearance as wisps of saturated air rose slowly, and multicoloured patches of dead leaves on the ground were spotlighted by sparkling sunbeams and seemed to be invested with some mysterious importance.

The sun bathed the tree boughs in golden morning light and it was possible to see the surprising variety of bark colours to great effect. There were pitted oranges and flaky reds, smooth greys, gnarled blacks, furrowed browns and mossy greens. The leaves represented every shade of green imaginable, but many of the young leaves were bright red and others were white, hanging from the branches like rows of fresh-filleted fish. All the colours of the trees against the dark blue of the early morning sky presented a glorious picture lacking only the desired presence of a *beelow* group in my view.

I went northwards towards a fruiting fig tree I had noticed a few days before. It lay just beyond the limits of the area I knew Sam and his family used and I hoped that they might venture that extra bit further to feed in it. When I reached the fig tree it was full of birds. There were several species of vociferous green and pink pigeons, dazzling golden orioles, hill mynahs, glossy starlings with blood-red eyes and, most striking of all, a green broadbill. This bird is no longer than a European starling, though plumper and squatter. In outline and flight it could conceivably be mistaken for a small owl but there the resemblance ceases. The male is a deep, glossy, iridescent green with large dark eyes, black bands on its wings and a bold black spot on either side of its neck. The colour of this species as a whole is generally considered to be particularly vivid, but the form found on Siberut is distinguished by being even more brilliant than its cousins on the mainland. Due to its relatively small size it is more often heard than seen and the local name for it, *luitluit*, amply describes its rather bell-like notes.

The branches of the tree were covered in pea-sized figs growing singly along their lengths. They seemed to be at all stages of ripeness, from unripe green through bright yellow and orange to over-ripe crimson. Fig plants, which can be small trees, large trees, lianas, stranglers or epiphytes, characteristically bear a great profusion of figs at roughly regular and often frequent intervals. Figs are not fruit in the strict sense of the word because their development is quite different, but they provide many of the fruit-eaters in the forest with an abundant, nutritious food over much of the year. Their life history is particularly interesting; just before the fig is fully formed, a minute female wasp crawls through a small hole at the top of the fig and lays eggs in the base of some of the many little flowers inside, pollinating many others meanwhile. By the time she has finished, the entrance hole has closed completely and she is slowly crushed to death by the growing fig. The wingless males hatch first and they visit the flowers where their sisters will hatch, ready to mate with them. Together the males cut a hole through the wall of the fig by which time the winged females have become thoroughly covered in pollen from wandering around inside. The females leave by the males' exit hole and fly on to pollinate another fig. Thus, figs and wasps are totally interdependent, and judging by the number of figs around, this amazing drama in miniature must be enacted almost every day.

The scores of birds above me were continually changing position, dislodging showers of ripe fruit from the branches. As I sat and watched the birds, I became aware that one of the chucking sounds I had associated with them was, in fact, coming from the ground. I craned to see what was making it without making myself too conspicuous and saw a small dark brown shape scuffling around among the 'dog's mercury' plants. The chucks were now followed by a high ascending *'wheeeee'* and as each chuck was uttered so I saw a sharp movement. Then the shape emerged and showed itself to be a *soksak* ground squirrel; each time it chucked, its short tail cocked and its head gave an upwards nod. After his success with the *jirit* Stumpy had proudly brought a young *soksak* into camp and we had managed to take it from him before any physical damage was done. In contrast to this adult, the young one had had three bold stripes down its back, similar to many other young animals such as tapirs and wild pigs. Interestingly, a close relative of the *soksak* on Sumatra has stripes when adult, and it is possible that some ancestor of the *soksak* gave rise to the Sumatran ground squirrel, keeping some of the *soksak* juvenile characters. If this were so it would be an example of the evolutionary phenomenon of 'neoteny' by which an 'advanced' form of animal tends to retain juvenile characteristics of the more 'primitive' form. Other examples include the pink skin, flat face and sparse hair of an infant chimpanzee having a resemblance to ourselves, and the fish-like gills possessed temporarily by all mammal embryos.

Oblivious of its evolutionary interest, the *soksak* busied itself inspecting the fallen figs. It disregarded the very ripe ones which had probably started to ferment, but took small bites from others, continuing to chuck and nibble until a noisy flapping of wings and loud rasping trumpet-like calls sent it scurrying into the undergrowth again. Hundreds of figs hit the ground sounding like a shower of hailstones and when I looked up I saw heavily swinging branches supporting a pair of pied hornbills. As hornbills go this species isn't particularly large, being only two feet long, but it was one of the largest birds in the Siberut forests and always exciting to see. In Sumatra, seven or more hornbill species may live in an area but on Siberut there was only this one species. This illustrates the general paucity of animal (and probably plant) species on Siberut which could be predicted by virtue of its relatively small size and isolation. Similarly, Britain, on the extreme west of Europe, has far fewer animal and plant species than the continent. Siberut would be expected to support only about one-quarter of the number of species present on Sumatra and this has meant that whole groups of birds, such as pheasants, tits and woodpeckers, ubiquitous on the mainland, are absent.

These pied hornbills continued their raucous calls for a while and then began to feed. They picked figs delicately from the branches with their huge cream-coloured bills, seemed to palpate them briefly and then jerked their heads back so that the figs dropped neatly down their throats. Few of the other birds stayed on the boughs the hornbills were using, probably because they would get too rough a ride when the large birds hopped along the branches to find more figs.

Sam hiding

Behind me there was a crash of leaves from the north-east; then another, and another. I crouched down peering cautiously towards the source of the sounds. Sure enough, the female and juvenile *beelow* were approaching the fig tree as I had hoped, and about ten yards behind them was Sam. At supper last night I had decided to name the female Bess and had named the juvenile after Katy Chivers, my Cambridge supervisor's charming toddler. Sam often travelled behind the others, apparently content to let Bess and Katy take the lead. My view of them wasn't yet very clear but I consoled myself with the thought that at least they probably couldn't see me any better. I waited until they were all in the fig tree and then broke cover slowly, inspecting leaves as I went and not looking up at all. It took only a few seconds for one of them to see me and immediately it gave the quiet alarm hoo. All three of them then hooed at me but I refrained from looking up and continued to pick leaves, chewing on some that looked young and tender, until the hooing had subsided. I then turned round and walked back, but this time looking up furtively. As soon as I made eye contact with Bess she hooed again; I cast my eyes downwards and the hoos ceased.

The *beelow* seemed to be torn between fleeing and gorging themselves on figs. They could eat a large number of figs very quickly because, in contrast to many fruit, these could be removed in handfuls and eaten whole. Figs were therefore a favoured food and unless I did anything really frightening they would probably eat until full. I walked slowly back and forth beneath them, making occasional brief eye-contact. At some time in the future I would have to make detailed notes of their feeding rates and of the use they made of the tree crown, but for the time being I was content to let them do the observing. In some areas of Siberut the people are often heard to say *'moile, moile'* or 'slowly does it', and that was how I hoped to gain the confidence of these *beelow*.

With no warning, Katy leapt out of the fig tree and the others soon followed. I picked up my bag and metaphorically girded my loins ready for a chase through the forest like the other day. I stood, poised for action, as though waiting for a starter's whistle, but nothing happened. I put my shoulder bag down again but held onto the strap, not wanting to be caught out. Still nothing happened. I crept very quietly up the path to see if perhaps I had lost them. No, there were Bess and Katy sitting close together, hunched over with their eyes shut. I moved out of their view again, sat down and waited. I felt it would be unfair to frighten them after their large meal for they might never forgive me if they were chased and made to suffer the agonies of acute indigestion.

While they rested I wrote notes, drew some sketches and read an Indonesian-English dictionary, looking for useful words. Every half hour I went forward to check that the *beelow* were still there and found them in varied poses of abandon, sometimes in the shade and sometimes in the midday sun. Sam gave away his position when the rhythmic movements of his arm scratching his armpit caught my eye. He saw me too, frowned, pouted and hooed very quietly but none of them moved. After two hours of being in the same position I was beginning to feel quite pleased with myself, although for the majority of the time I was collecting as much data as Jane had when her *jirit* had stayed motionless the other morning. Jane would probably never know if she saw that individual again, although some had characteristically long or short tails. In contrast, every minute I could stay near the *beelow* reduced the time necessary before they would accept me as part of the forest scene.

As abruptly as they had ceased feeding, Katy and Bess suddenly leapt out of the tree and rushed back along the route they had used to enter the fig tree. As usual, Sam seemed to wake up to what was happening after some delay; he scratched himself some more and followed. I maintained a discreet distance hoping not to worry them, but failed; I had walked barely twenty yards when Katy caught sight of me and hooed sharply. It needs only one nervy gibbon to make a nervy gibbon group and they all fled up the hill. I had no paths in this area yet and so had to scramble after them. Some parts of the hill were nearly vertical and I had to climb on tree roots and liana stems to keep up. Even the *beelow* appeared to be travelling with a modicum of difficulty but at least they managed to maintain a certain standard of grace. They are wonderfully adapted to travelling through the trees. They can walk upright on their back legs or on all four limbs, they can leap and, most characteristically of all, they 'brachiate' – swinging hand over hand incorporating coordinated movements of the legs, like a child on a swing. So accomplished are they that there are sometimes moments when a brachiating gibbon isn't touching any part of a tree at all. Gibbons' arms are exceptionally long in proportion to their body length, and when the distance they travel through the air with no support is considered, each 'stride' is extremely long. In addition, brachiation uses very little energy and so allows gibbons to travel quite long distances each day.

I was covered with clayey mud that was now running in sweaty streaks across my skin after the exertion of the hurried climb. When I eventually reached the top of the slope I couldn't see anything moving above me. I rested on a cool, soft, mossy log and no sooner had I begun to catch my breath than I saw the *beelow* travelling down another ridge away from me. They must have moved much faster than I had imagined.

I got up and raced after them, digging the heels of my boots into the soft mud of the steep ridge path to save myself from slipping. Three minutes later I was beneath the trees in which I had seen the *beelow* from the top of the slope but now there was no sign of them. I stood and waited, calling to them every so often hoping they might give themselves away with a nervous hoo if they were within earshot. I ventured down a few nearby paths but eventually decided to cut my losses and to clear paths along the new routes they had used during the last few days.

Of these paths, the one across the swamp presented the greatest problems. I started over on the West Ridge side and the first task was to cut down a small tree as a means of getting across the sluggish, ale-coloured Situi river. I cut a ten-foot pole to provide myself with extra support, and with that stuck firmly into the mud below the water, I shuffled my feet gingerly along the four-inch-wide bridge. It lurched precariously just above the water as I grappled with rattans, ant-covered branches and large needle-coated arum leaves that were in my way. Soon, however, I reached a nest of tangled roots that I could stand on at the far end of the bridge, and from there I picked out the easiest looking route in the appropriate direction. The peaty ground was covered with a thin layer of water but it felt firm enough just below the surface. I tried standing on it, first with one foot, and then with both. Fine. I had a slight sensation of sinking but not enough to worry about; then the matted crust of roots and half-decayed vegetation gave way and I was up to my armpits in acid, swampy ooze. I couldn't find anything firm and struggled through the tough but yielding peat for a few minutes before I managed to grab a small tree and haul myself onto the rough roots around its base. I leant against the tree for a moment or two and then tried again, this time taking shallow bounds rather than stopping to ponder on the instability of the soil. I reached a fallen tree, climbed onto it and on looking round saw the surface of the swamp throbbing and pulsating after my hurried footsteps. With all that behind me, I checked the compass bearing and continued to clear a route sufficiently wide for me to run along after my *beelow*. The swamp forest here was quite different from that on the hill; the assemblage of species was totally distinct and all the trees were noticeably shorter than elsewhere. Frozen cascades of knobbly roots fell from many of the trunks on either side of me, and roots of others sprang out of the stifling soil at intervals in a desperate bid for air. There were few shrubs or herbs but some elegant fish-tail palms formed single-species woods of their own where a tree had once fallen and decayed.

Over the next three days I had to use that ghastly swamp path twice to follow Sam and his family, and each day I spent longer with them. They allowed me to watch them feeding as they had done in the fig tree but moving between feeding trees was still rather chaotic with either Bess or Katy getting nervous and tearing away from me. The key was, however, that they used about six sources of fruit more or less every day and when I lost them, I could often relocate them by going to the other food sources. Each bout of feeding only lasted about fifteen minutes but for some time afterwards they often rested nearby.

One morning in the following week Sam started to sing at about four o'clock while I was still in camp. I was beneath his sleeping tree twenty minutes later and managed to record most of his song. It started with soft piping notes each separated from the last by as much as a minute. The pipes became longer, like long descending whistles, and then two or three whistles were joined together. Progressive elaborations of this theme followed until fast ascending whistles became a short trill. Each phrase was not necessarily the same as, or more complicated than, the preceding phrases because Sam often regressed to earlier phrases and began anew. Even when he reached the trill phrases, he rarely gave more than two in a row. The song could be heard nearly a mile away and was really quite loud when one was close to him. Despite this, Bess and Katy appeared to continue sleeping in the top of a neighbouring tree, Katy nestled in Bess's arms. How they were able to sleep through the din defeated me.

The next morning I walked briskly towards the 'Glade' where I suspected the *beelow* might be sleeping. The torchlight was trained on the ground some feet in front of me and I watched the leaves, dead branches and roots pass under my feet. Then I noticed one particularly large, green leaf come into view. Within the fraction of a second between then and the time I reached it, I thought to myself that I hadn't noticed such a large, strangely-shaped leaf before. The contours caused by the leaf venation gave the peculiar impression that it was coiled up. It also seemed to have two free ends, the thicker one pointing towards me. It took rather too many milliseconds for the cells in my brain to switch from 'leaf recognition' to 'other object recognition', and, by the time the word 'snake' came up as a flashing red light behind my eyes, I was standing with my right foot on a writhing pit viper. I immediately shone the torch onto my foot and watched, aghast, as the pit viper held its mouth wide open and tried to bend its three free inches of neck round to make contact with my boot. The two translucent curved fangs were all too clearly visible and looked quite capable of penetrating the canvas of my boot without any great effort. The textbooks tell you that you need have little to fear from snakes in the forest for they sense your footsteps and slither silently out of the way; that's the textbook version.

I figured I had two choices: one was to race ahead and hope the snake wouldn't make a lunge at my bare calf as I started, and the other was to try to kill it. The deadly venom of pit vipers is used medically in the treatment of thrombosis but only in small, finely-measured amounts. Since I didn't have a thrombosis I decided on the latter course but, unfortunately, I hadn't brought my *parang*. As I shone the torch around me looking for some kind of weapon, I realized that most of the plants within arm's reach were delicate ferns and shrubs which would have been as much use as a pea-shooter against an elephant. There was, however, a small sapling which I reached by holding onto a shrub with my right hand to balance myself, bending my right knee, hooking the toe of my left foot around the sapling, drawing it slowly towards me and catching hold of it with my left hand. I broke off the top at the thickest part possible and, standing like a model for a statue of Eros, tore off its branches. Using my teeth and fingernails I urgently fashioned a very rough chisel-point from the wood and managed to position this on top of the ever-extending neck of the snake. I pushed down hard and stepped away cautiously. The head end writhed but the tail end lay trembling, suggesting that I had successfully managed to sever the backbone. Snakes take a horribly long time to die but eventually it lay still. I felt remorseful when I saw its beautifully bright grass-green silky skin, its red tail, the delicate white dots along its sides and its yellow belly, but there would have been little chance of my feeling anything again if I had been bitten. It wasn't wasted, for several months later it was received gratefully by the Natural History Museum in London.

Chapter Three
Tale of Two Turtles

Jane was undoing the mud-encrusted laces of her boots after a fruitless day spent looking for squirrels when she called to me inside the house 'There are two children coming; I think it's Aman Mynoan's young boys.' The elder of them was holding a plant in his hand, presumably a response to our request for any orchids they found on fallen trees.

'*Anaileuita aleh*,' I greeted them from the veranda as they came along the narrow path from their father's *sapo* upstream, a fair bellow's distance away.

'*Anaileuita*,' the elder brother replied. '*Maobakkaap angrekta*, do you want any orchids?'

'Let's have a look at them,' said Jane. He held out what looked predominantly like a bedraggled bunch of green basset-hound ears. 'Oh, that's lovely. Look, a beautiful flowering *Phalanopsis* orchid,' she exclaimed as she held the inch-round, purple and yellow flower between her fingers. She sniffed it, and from the expression on her face it obviously had a wonderful perfume. Jane fetched an empty bottle with a good screw-top for payment, and the boys obviously felt they had the better half of the bargain.

'How did you find the orchid?' I asked.

'Do you remember that strong wind this afternoon just before it rained?' the elder boy said. I nodded. 'Well, an old durian tree fell over near the *sapo*. *Teelay*, it made a loud noise! The orchid came off the top of that tree. Do you want any firewood? Durian burns very well.'

'Indeed we do. Tell your father we'd be glad to buy some.' I went inside the house and tore some tobacco off a block for the children to give their father.

'*Kawat*,' I said, thinking the brief transactions at an end for the day.

The 'spokesman' looked at his younger brother as if to say 'go on, now you'. The brother looked shy and his eyes pleaded for the other to continue speaking.

'*Teelay*,' the boy said in quiet exasperation. '*Maobakkaap nene*,

37

do you want this? It was in a hole in the durian tree.' He looked at his brother indicating that that was enough of a cue, and the little boy brought a sticky hand from behind his back to reveal a furry grey shape nestling comfortably within it.

'*Apaonynia*, what's that?' asked Jane excitedly.

'It's a baby *peeow*,' he replied.

'What's a *peeow*?' asked Alan, who had just come over after emptying the rain gauge.

Peeow in hand

'*Teelay*!' the boys said to each other, staggered by our ignorance, 'a *peeow* is a type of squirrel that comes out at night.'

As they finished speaking, the *peeow* lifted its fluffy round head and peered at us, bemused, its fur all tousled as though it had only just got out of bed. Its mouth twitched and then it broke into a broad and obviously very satisfying yawn.

'We'll take it,' I said, and neither Jane nor Alan appeared to be horrified at the prospect of adopting the squirrel. I went into the hut again and found a hundred rupiahs to give the boys and some more tobacco for Aman Mynoan. Jane took the *peeow* and the boys took their leave.

'Now what?' Jane said. What indeed. Alan and Jane had built a stout cage from old planks and chicken wire in the hope of keeping a couple of *jirit* squirrels for closer observation and that was obviously going to be the *peeow*'s accommodation for the time being. While Alan fetched some paper padding from the top of a tin of biscuits for use as nesting material, I put on my boots and went out to find some forest fruit before nightfall. By the time I returned, Peeow, as he had by then been named, had been wrapped up in the padding and Alan had positioned stout twigs in the cage. Cooked rice, ripe bananas, some young leaves from a small tree in the clearing and a little tinned meat were already in the bottom of the cage and to these I added my offering of selected fruits. We fastened a heavy sack over the front of the cage and resolved not to look inside until morning.

The cursed alarm clock rang as usual at 3.30 am and Jane switched it off for me. I could hear Peeow walking around on the floor of the cage and so went over quietly and shone a shielded torch onto the food. He had rearranged much of it, but the only sign of feeding was a patch of double-teeth marks on the ripe banana. Peeow came up and looked at me but I didn't hold his attention for long, and he returned to his paper nest. Jane poked her head out from beneath the mosquito net to ask how Peeow was and, since he had scarcely eaten, we decided to feed him by hand on some mashed banana. We weren't sure just how young he was but he could well still be suckling if the durian tree hadn't fallen over, and he might not be used to solid foods yet.

I fetched Peeow out of his cage, receiving only a few nips from his budding teeth in the process, and carried him over to the bed which was now lit by a lamp at its side. We tucked the net under the mattress just in case he felt flighty, but nothing could have been further from his thoughts. Instead, although alert and wide awake, Peeow moved only slowly and deliberately from me to Jane and back again, exploring the curious new world. We mashed bits of banana between our fingers and, after only a very little persuasion, Peeow lapped up the pulp. He ate what we regarded as a prodigious quantity for a young flying squirrel but took his time about it, and it was dawn before I could put him back in his cage. Once covered with shredded paper he curled up, holding his bushy tail over his face, and fell asleep.

Peeow became bolder every day and it was a delight to have him in camp with us. Jane and I fed him at intervals through the night, taking it in turns to get out of bed and in three weeks he'd doubled his weight to four ounces. Feeds took longer and longer as he learnt the enormous potential for exploration that our work table provided. He even encouraged us to play with him – running up to the nearest hand and patting the wrist. He leapt away, his tail describing graceful curves in the air, and waited to be patted back. Then it was his turn to pat us again. All this was very sweet and charming but at two in the morning I'd really much rather be sleeping.

On the south side of the clearing there stood two mature coconut trees that never bore any fruit. Aman Bulit and others told us that this was because flying squirrels repeatedly visited the trees and ate the very young fruit. Their damage was sufficiently serious in some of the western river basins whose inhabitants depended on selling coconut meat for an income, for communal hunts to be organized to shoot out the squirrels from an area. Occasionally we took our brightest torch into the clearing at night and watched adult *peeows* gliding with amazing accuracy down from the forest edge to the coconut trunks thirty feet away, feeding on the developing fruits, and then gliding back again. Although there is a single published record of a flying squirrel actually flapping its 'wings' and ascending, it seems highly improbable. Flying squirrels glide by stretching flaps of skin that join their ankles to rods of cartilage at the side of the wrist that are held out at right angles to increase the effective surface area of the 'wing'.

After Peeow had been with us a month, I cut a sapling from the forest and wedged it firmly between some floorboards to encourage him to practise gliding. Initially he was rather unsure of exactly what he was supposed to be doing, but at length he climbed delicately to the wobbling end of a branch, peered forward with intense concentration at Jane standing two feet away, and wiggled his hindquarters like a cat about to pounce on an unsuspecting prey. Then he leapt with arms and legs akimbo and landed clumsily but successfully on her chest and scampered up to his usual perch on her shoulder, shooting lascivious glances at her tasty-looking earlobes.

There came a point, a month later, when we felt Peeow should be released back into the forest. A few weeks previously we had been seriously concerned about any chance of his survival in or out of the forest, when Stumpy had crept unnoticed into the house while we were feeding Peeow and pounced on him. They were both under the table before we realized what was happening and by the time Jane had prised Peeow from Stumpy's mouth, Peeow's shoulder and back were bleeding and he seemed to be in shock. He was listless and ate very little for several days, seeming unable to use one of his front legs, but only a week later his behaviour was almost indistinguishable from that before his traumatic experience. He was at his most endearing when he was dopey at dusk just after the first feed of the day. He would lie on his back in the palm of your hand, eyelids half-open, licking sticky banana absent-mindedly from the soles of his feet. Then he would rest his head on your fingers, shiver slightly and launch into a wide, wide yawn. He remained happily on his back like that, soliciting for his soft grey belly to be stroked, until darkness had set in and he became more active.

We fed Peeow, as usual, at dusk on the day of his release and then all went into the forest along the Flat Path to a place where I had recently heard chatters, chuckles and *'treech'* calls from adult *peeows* and where a fig vine was heavily loaded with ripe, purple fruit. For the past fortnight Peeow had greedily accepted figs and other forest fruits

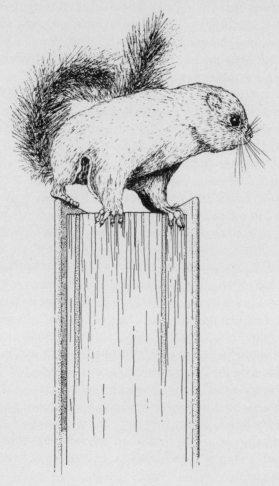

Peeow

and we were confident that he would now be able to look after himself. He had ridden on Jane's shoulder as we walked and no sooner had she put him on the large tree encased by spiralling vine stems than he started exploring. He nibbled at bark, nosed under dead leaves, chewed at moss and scrambled ever upwards. He never looked back; but we missed him and his charming ways.

After another morning of seeing nothing but trees, I came across Jane resting at Summit Two after her intensive systematic walks around the study area during the three hours after dawn when the squirrels seemed most active. We consoled each other on our common lack of any startling success, but she was able to teach me to recognize a distinctive rising trill she'd just identified as coming from a *jirit*. I'd now be able to make notes when I heard it because, although watching *beelow* occupied much more time each day now than I'd have thought possible a month before, there were still plenty of hours in which I was able to make notes on other aspects of the forest.

'Do you want to meet here for lunch?' I asked.

'If we can assume that I won't be too busy making voluminous notes on any courting *jirit* or *logas* making nests,' she said dryly, 'then I reckon I can make it.'

'Me too. One o'clock?'

'One o'clock.' We went our separate ways with me deciding to walk around the north of the area I knew Sam and his family used. Half an hour later I heard crashing leaves in front of me. Great, I thought, *beelow*! I chased off in the direction of the swaying branches and saw an indistinct shape move away northwards; I followed. Its movements somehow looked wrong for a *beelow* and when I saw that it had a long tail, I realized it had to be a *joja* monkey. Its black and grey body streaked ahead, its bouncing, startlingly white scrotum identifying it as a male, but now and again it turned, sat facing me and gave a loud, repetitive, rapid '*bagok*' call. It raced on, taking me beyond the limits of my previous explorations, to a ridge where the red-barked dipterocarp trees were huge and the ground was covered with squat, prickly palms through which I half cut, and half pushed my way. The *joja* was getting further and further in front and then I saw him make an abrupt right turn. I reached the tree in which I had last seen him and cut out a bold cross on its trunk so that it could be incorporated on a map at some time in the future. There was no longer any sign of the *joja*.

I returned to the centre of the high, wide ridge to find the route by which I had arrived. I started downhill but found that the path soon petered out. I pushed to the left and found another path of sorts but that meandered about and after a while petered out too. I pushed to the right but with the same worrying result. To reason out the situation, I sat down in a small clearing and stuck the tip of my parang at an angle into an old log, drawing around its shadow with a sharpened twig. The sun was virtually overhead but after about fifteen minutes it had dropped sufficiently for me to work out from the new shadow where the points of the compass were. Camp lay to the south or south-west but none of the ridges around me seemed to run in that direction. I followed the most hopeful-looking one.

Walking down the small ridge-top wasn't as easy as I had expected. I cut down to the stream that I could hear below me and walked along it, its slippery sides getting darker and steeper. Where I couldn't wade in the water itself, I 'brachiated' between small tree trunks rooted precariously at the sides; it made me appreciate the competent skill of Sam and Bess. Ahead, I could see the stream take a plunge and when I reached the lip, I found myself looking down a sheer waterfall. With the sides now far too steep to climb, I had little choice but to stand on the slippery edge, hold my nose, and jump. The forest rushed past my eyes and after what seemed like an excessively long time, I plummeted into the clear, turquoise pool beneath. It was luckily quite deep and surprisingly warm; after checking I was still in one piece and shaking the water out of my eyes, I trod water for a little while, looking up at the magnificently delicate, trellised vaulting of the canopy. I swam with the water's flow and saw that I would soon be joining a larger river. I scrambled out of the water and clambered over a spur, brushing my leg against some innocuous-looking leaves. Appearance can be deceptive in tropical rain forest, however, because it must have been a close relative of the *alalatek* nettle that had stung the three of us on Christmas Eve. By the time I reached the main river, the glands in my groin, armpits and neck felt the size of footballs and I'm convinced I must have looked like a Michelin man.

Until then I had thought that I was in the Paitan catchment area but the shallowness of this river and the formation of the gravel and mud was different. I carried on along its sweeping meanders because the settlement pattern on Siberut is such that if you walk down almost any river you'll eventually find people; even if you have to reach the coast. An hour later I heard distant voices and at length encountered Badi, a teenager from Sirisurak who had helped us in the past to get stores up

43

to camp, and a couple of his friends. All they seemed able to say when they first saw me walking down the river was '*teelay*', but after recovering from their shock we exchanged greetings and news. They told me we weren't standing in the Paitan river but in the Mauku river to the south-east. Badi suggested that the easiest way home was via Sirisurak and so we set off for the village at a jog trot through the forest.

Aman Bulit was at home, sitting on a bench on his veranda with his old Rolex Oyster watch (a present to him from Reimar Schefold) in one hand and his two-foot *parang* in the other, effecting repairs.

'*Anaileuita, aleh*,' he welcomed me with a big smile, 'what are you doing here? Do not tell me you are lost!'

'How did you know?' I asked.

'I will tell you later, *aleh*.' He went inside and stopped Bai Bulit sifting sago flour for their supper and told her to heat up some water. I sat gratefully on one of the other rough benches and about twenty people, including the village chief, gathered to ask why I'd arrived unexpectedly in the village. I told my tale about the *joja*, feeling rather shame-faced and foolish as I did so.

The village chief came up to me and said, '*Tak leu joja ka edda; kupaatu sikatengahloinak*.' He didn't believe it had been a *joja* that had led me astray but a creature called *si-ka-tengah-loinak* or 'the one that lives in the middle canopy'. 'Ah, but it looks like a *joja* at first glance,' the chief continued in his sage-like fashion as Bai Bulit brought out cups of sweet hot coffee, 'the difference is that the middle of its belly and back are white. It distracts people in the forest and makes them lose their way. It happened once to my grandfather,' he insisted, and the skinny old men around me nodded their heads as they remembered the event; 'he was out looking for suitable trees for dugouts when he saw a *sikatengahloinak* and followed it – just like you did. It was three days before he found his way home again.'

This seemed to me like a magnificent face-saving way of disguising the fact that I had not been able to find my way back to camp. I therefore agreed wholeheartedly that a *sikatengahloinak* was probably to blame. The old men reminisced and Bai Bulit brought out some deep-fried bananas for the assembled company. Whilst sharing a plate of these with Aman Bulit I asked him quietly whether he would take me back to camp through the forest because the sun was getting low and I couldn't risk getting even slightly lost, particularly if the ridge were inhabited by a *sikatengahloinak*.

A few minutes later we waved goodbye to the men of the village and, pretending not to be too tired, I tried to keep up with Aman Bulit's rapid pace. During a short rest at the top of a hill called Teitei Jotjot I asked him about the round patches of dead 'dog's mercury' I had seen in the study area.

'*Teelay*, you have not been too close to them, have you?' he asked.

'Well, yes, I suppose I have.'

'You must be careful!' Aman Bulit looked serious.

'But why?' I questioned, 'is there some kind of poison around them?'

'In a way. Look, those patches are made by *sikaseddet*, the otter. You must never go looking for it. *Sikaseddet* has a really hot *bajo* and when it sleeps in the forest, the plants around it shrivel and die of fever.' Aman Bulit looked deep into my eyes; he wasn't joking.

'What exactly is a *bajo*?' I asked. 'I mean, how do we recognize it?'

'You cannot see or hear *bajo* but you can sometimes feel it. Everything has a *bajo*; the animals and plants, you and me, rocks and the sea. You do not need to be scared of *bajo* but it is as well for you to know about it. Some things such as otters, *umas* that have just been newly dedicated after days of ceremonies, and old *sikereis* all have a strong *bajo*. The fact that their *bajo* is stronger is not necessarily good or bad, but if you come close to a strong *bajo* too quickly it can hurt your soul and that will cause you to get a fever.'

'So, it's the soul that gets affected by a strong *bajo*.'

'That is so,' Aman Bulit replied. 'If, when you are asleep, your soul goes wandering off, it might linger around a burial place or it might gaze too long at a rainbow. If the soul is reached by their strong *bajo* it gets frightened, influences the body and that is when you become ill.'

I paused. 'Can you die?'

'Yes,' Aman Bulit replied slowly, 'you can die. The soul feels ill too and rushes to the ancestors. Then,' he looked at the ground, 'then the soul might be fed and clothed by the ancestors and not wish to return to your body. If this happens you have to die. It is possible to cool a strong *bajo* but only a *sikerei* knows how. So be careful.'

As we walked on hurriedly, he told me of a dream that had troubled him the night before and which had led him to suspect I had been lost when I arrived in Sirisurak. He dreamt he had been fishing for turtles off the east coast of Siberut. After many unsuccessful attempts he caught two of them but they had managed to escape from his dugout. He had searched the seas for a long time and just as he was

45

beginning to despair, he found one of them again. For what seemed like days he had paddled on and only when he was starting towards home did he catch the second one. Then he woke up.

After two hours' walking we arrived at camp and immediately realized the coincidental significance of the dream — Jane wasn't there. Alan was bent over the fire cooking supper, expecting us to roll home together at any time, and he confessed he was just beginning to get worried. It was always possible that she'd been unable to leave lots of exciting squirrels but, seeing as how she'd have needed floodlights to see anything by now, this seemed extremely unlikely. We found flash-lights for the three of us and we went out into the pitch-black forest with some hastily assembled morphine, splints and bandages, yelling out to Jane as we walked. We stood stock still now and again, trying to hear any sound she might be making but with no luck. At Summit One, a long-tailed giant rat ran hell-for-leather across our path and only yards further on, I bumped into Alan's back as he bumped into the heavy shoulders of Aman Bulit. Directly ahead was a confused, plump, eight-foot, mottled-grey snake, its raised head swaying from side to side. Aman Bulit rustled some leaves and shouted at it. We watched silently as the entire length of the snake glided elegantly into the base of a clump of *ariribuk* — just where we might have built a rain shelter. We built no more rain-shelters.

At Summit Two we found Jane's shoulder bag and her packed lunch. We walked around the summit and down some of the nearby paths, calling intermittently. But still there was no sign of her. We returned home along a different path and our hopes that she might by then have reached camp by yet another alternative route were not realized. We ate a rather muted supper while we pondered on all the fates, good and bad, that could have befallen Jane in the forest. When I went across the clearing to our house, Aman Bulit put his warm arm around me and said, 'Do not worry yourself. God will be with Jane wherever she is and will be looking after her. Pray for her.' I did. The word he had used for God was '*Taikamanua*' or 'spirits of the air'. This was adopted by the early missionaries on Siberut as a word to embrace the reality of God, and when you live in the forest it is a very apt description. I slept in fits and starts, not helped by having to get up and feed Peeow every four hours.

Meanwhile, Jane was shivering on a sandy river bank swatting countless mosquitos as she watched the moon disappear from sight below the forested hill in front of her. She had got lost too, by

46

following my rapidly cut path to the north in the hope of finding me with some *beelow*. She had had precisely the same trouble as I had in trying to find a path or a ridge leading back whence she had come. One of her floundering peregrinations took her round in a large circle during which she found a huge landslip down a steep slope that provided a rare view across rows of forested ridges that must have included a fair proportion of the Saibi basin. Jane had been sure I was somewhere on the high ridge and called out to me; but by then I had been well on the way to the Mauku river.

Jane's descent to a main river was as difficult as mine and she kept repeating to herself that, as long as she did nothing stupid like breaking an ankle, she'd be sure of finding someone eventually.

At dawn the next day, Aman Bulit went upstream to summon help for a thorough search from members of the Sagaragara clan. Alan and I left for the forest with our medical kit and headed for Summit Two where we had found Jane's bag and lunch. From there we searched round in ever-increasing circles but could find no sign of her. Then, almost three hours later, we heard Aman Bulit yelling our names. He found us and related how he had decided to explore one of the Paitan's tributaries in the headwaters before getting help and had happened to find Jane coming downsteam. We all ran back to camp and found her slightly the worse for wear with badly scratched legs, but so pleased to be home that she couldn't stop talking about the waterfalls, an eight-inch, turquoise and red scorpion, the enthusiastic leeches, the clinging rattans. . . .

Supper that night was a happy event. The four of us sat on the veranda afterwards, chatting and sipping coffee, watched all the while by Peeow chattering conversationally from his cage.

'Has anyone ever told you the story about the python and the *simakobu* monkeys?' asked Aman Bulit.

'No, I don't think so,' Jane replied. 'Could we hear it now?' knowing full well that was why he'd brought up the subject.

'Very well, I will tell you.' He cleared his throat as though preparing to make a formal speech and then began. 'In former times, there lived in the depths of the forest a large and hungry python. One day he was slithering through the tops of the tall trees when he spied a family of *simakobu* monkeys some way away. Very slowly, but very surely, he approached them; closer and closer, closer and closer. As soon as he was able, he darted forward, strangled one of the monkeys and let it fall to the ground. Thungk. Then he caught another, and another and

47

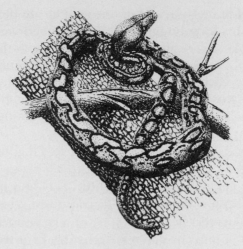

Python

yet another, until he had dropped all four of them to the ground. The python did this because, had he eaten the monkeys in the tree, he would have been too heavy to climb down again.

'Now, although there were usually very few men in the forest, there happened to be two brothers passing by and they heard the noise of the falling monkeys. They guessed what was happening and so they waited until all four monkeys lay dead on the forest floor before grabbing them and rushing off into the forest. When they thought they had run far enough, they stopped to make a fire over which to dry the monkey meat. When night fell, they slept by the fire for warmth and to guard the meat.' Aman Bulit shifted his position on the chair because one of its legs was disappearing between the floorboards. He regained his dignity and the thread of the story and continued.

'Meanwhile, the python had reached the ground and found the smell of his dead *simakobu* was mixed with that of the two brothers. He slid steadily through the forest, tracing their scent, until at last he came to the brothers' camp. He circled round and round, round and round the sleeping bodies, hissing as he went, each time getting closer to where they lay. The younger brother stirred and heard the python. He shook his brother to wake him but, try as he might, could not. The python was circling closer and closer, closer and closer, until the younger brother could stand it no longer and ran away to his father's house.

'The python circled closer still, with one eye on the *simakobu* meat

48

above the smouldering fire, and in one gulp he ate the elder brother. He then curled his long body around the fire and went to sleep.'

Jane was looking very glum; presumably she was thinking that she could conceivably have been a python's supper the night before. We had remarked to each other some weeks before that it was only the python we need worry about while sitting still watching animals. We had consoled each other then, however, with the thought that at least pythons were largely nocturnal.

'Is that the end?' Jane asked, 'because if so, I'll go and make us all some more coffee.'

'Oh no, not yet,' Aman Bulit replied. 'Stay and hear the end; it finishes happily! A few hours later,' he continued, 'the brother inside the snake woke up and heard the song of a tailorbird. "*Teelay*," he said to himself, "it must be dawn but why can I not see?" He tried to turn but he still couldn't see. As he moved his left shoulder, so the dagger that he had strapped to his arm the night before cut open the side of the sleeping python and he saw the light. He cut more of the python's belly and squeezed out. He killed the snake and skinned it so that he could show people the size of the beast that had eaten him. His family was very impressed and after drums had been made from the python skin, they celebrated with a party.'

Aman Bulit rolled another cigarette and said, 'That is all. I did not scare you too much, Jane, surely?' he asked smiling.

'I'll go and make some coffee,' was all she said.

An hour later we were still swapping stories when suddenly the lamplight seemed to flicker and Stumpy leapt down from his habitual position on Alan's lap. None of us could make out what was happening but we voiced suggestions such as 'It's a bat' or 'It's an owl'. Stumpy was standing rigid on the table but we still couldn't see what had caused the disturbance. Following his steady gaze up to the underside of the thatched roof, I saw a huge female atlas moth resting just out of reach. This moth is a close contender for the title of 'largest moth in the world' with its wing span of nearly ten inches and a wing depth of little less. Its fat, furry body was about two inches long and the vast wings were patterned with subtle shades of brown. In both of its forewings was a transparent triangular window in which the minute scales never grew and through which we could just discern the lines of the thatch.

49

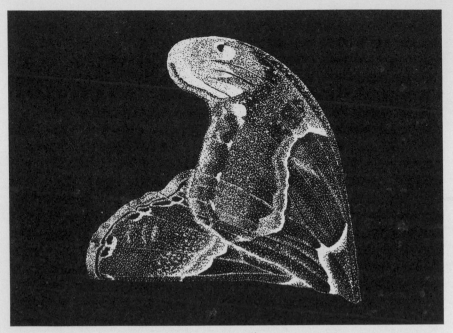

Wing of atlas moth

I built a tower of benches and tins beneath it while Alan fetched a camera and Jane tried to restrain Stumpy by taking him outside. Meanwhile, Aman Bulit continued to sit on the veranda but we guessed he'd have seen dozens of these moths before.

After we had photographed and admired it to our satisfaction, I picked it off the roof by the surprisingly rigid leading edges of its fore-wings. Aman Bulit told Alan to blow out the lamp and I launched the moth into the clearing. We sat chatting for a while and then Alan relit the lamp. Barely a minute later, Stumpy was leaping into the air again and the moth landed by the lamp.

'*Teelay*,' Aman Bulit exclaimed angrily, 'blow out the lamp and throw that thing away.' This brusque manner wasn't like him at all and we did what he said without question. Again we sat in the still darkness.

'Do you know what the name for that moth is on Siberut?' Aman Bulit asked. We didn't. 'It is *surat d'atay* which means "letter from the grave". You probably do not know yet how people traditionally bury their dead on Siberut, do you?' We shook our heads again.

'It is done this way. When someone dies, the clan members make a coffin. It is similar to a dugout only it is shorter and has a lid. The body is carried in this coffin up to the top of a high ridge, like Sabeuleleu or Teitei Jotjot, a long way from the clan's *uma*. The men build a platform on stilts and before they rest the coffin on this they make a drawing of a *surat d'atay* on the lid. Then they leave and never return to that place.'

'Why do they draw *surat d'atay* on the lid?' Jane asked.

'Because it is one of the ways the dead communicate with us. The dead man's soul can enter the drawing whenever it wishes and bring us a message.' He paused, deep in thought.

'What's the message?' Jane asked softly to relieve the tension.

'There is usually only one message; a member of the family is about to die,' Aman Bulit answered. 'I have only ever seen one *surat d'atay* before and that was a few days before a close relative of mine died.' (We learnt later that dead relatives are never named.)

Alan reached for the matches to relight the lamp but then thought better of it. We all went to bed using shielded torches for light.

Two days later, Aman Bulit's uncle died.

Chapter Four
Pekai

After eight consecutive days of going out into the forest before dawn, I felt it was time for a lie-in, at least until sunrise. Alan grabbed a handful of cream crackers before going out to continue his study of the survival of dipterocarp seedlings around their parent tree, but Jane and I decided to have a civilized breakfast of bread cooked in bamboo with the added protein of innumerable, tasteless, little black flour weevils. We sat on the veranda watching the long-billed spiderhunters and red-capped tailorbirds skulking in the bushes in front of the house, and slowly became aware of the unmistakable sound of a dugout being poled upstream towards us. After a few days of reasonably dry weather the Paitan was now quite low, as evidenced by the noise of a *parang* cutting through trees that had presumably blocked the river, and by the muffled cries of '*teelay*' from a man. We heard several more '*teelays*' as someone tried to clamber up the bank to the clearing. Aman Gogolay then appeared from behind the tall ginger lilies, greeted us with a loud '*Anaileuita*', and walked up the step-cut tree trunk onto the veranda, carrying his bow, quiver and ubiquitous *parang*. He was the oldest person living on the Paitan river and the *rimata* of the Sagaragara clan. A *rimata* has been likened to a 'master of ceremonies', and although clans don't have chiefs or headmen, the *rimata* is often regarded as the most senior clan member.

'*Anaileuita*', we replied and offered him a seat. He sat down rather unnaturally on the cane chair and searched in the folds of his loincloth for his tobacco tin.

'You are going into the forest today?' he asked.

'Yes,' replied Jane. 'We're making a picture of the forest by the Tolailai river using this.'

'*Teelay*,' he exclaimed, as Jane showed him the compass. He soon picked up the idea and was pointing with his chin towards river basins, villages and *umas*. Despite his apparent interest he seemed to be distracted by some thought. He was watching 'N' stay put as he moved the compass round and round but was doing so, we guessed, to avoid looking at us.

'Any tobacco?' he asked, having revealed that his tin was empty. Had we been bold enough, we could have found quantities secreted in other parts of his loincloth, but it didn't matter. He returned his gaze to the compass as he rolled himself a cigarette using strips of dried palm-leaf.

'You have heard the news?' he asked eventually.

'You mean about someone stealing Aman Taatnappe's biggest pig?' I offered.

'No, about the Pekai.'

'The what?'

'Not what, who. The Pekai, who carry big, hooked *parangs* and cut your throat. Who wait for you and then jump on you and take your head. I know someone who was out with his brother; he went off to investigate a noise he thought was a deer and when he came back his brother was lying there. No head. Dead. The Pekai they are no good; *tak maeruk, teelay*!'

Aman Gogolay was intent on telling us more gory details but despite the ever-increasing volume of his delivery and its ascending pitch, the vocabulary was becoming far too complicated for us. However, 'blood', 'dead', 'throat', 'cut' and 'kill' were all very frequent. We gathered that the Pekai were waiting in the forest everywhere, that we were to carry *parangs* with us wherever we went, and that there was no chance of his being our guide, as previously arranged, on the first of a series of surveys we were conducting for the World Wildlife Fund. We had been asked to survey the island in order to suggest boundaries for an effective reserve, and having local guides with us was absolutely essential.

So, the next day, Alan and I walked to Sirisurak to talk over our problems with Aman Bulit or Aman Taklabangan, his cousin, with whom we could speak Indonesian rather than Mentawaian. That morning the village was strangely quiet but we assumed everyone was in their gardens and fields. We found Aman Bulit's old aunt, Lasui, treading a pile of rice stalks to thresh them, and she told us that Aman Bulit was at the coast looking for land to grow clove trees on. Aman Taklabangan, however, was out in his paddy field. We knew vaguely where his fields were and so walked in that direction. It was nearly midday and the sun was directly overhead, shining through the bluest of blue skies. The fields were deserted except for sparrow-sized white-headed munias flitting between rice stems looking for ripe seed-heads and jet-black dragonflies hovering like minute jump-jets as we disturbed them by walking on the bouncy, mossy ground. The paddy fields occupied all the flat ground available, about one mile by half a mile, between two ridges one of which was covered with gardens and the other mostly with forest.

The intense heat was almost unbearable and the buzzing of invisible insects set our heads buzzing in sympathy. We called out to Aman Taklabangan again and again but heard nothing in reply. Giving one final full-blooded yell we started back towards the village to get a drink. We had walked barely three paces when we heard our names being called. We looked round and saw the figure of Aman Taklabangan, slim, muscular and wearing nothing save a small loincloth, outlined on the skyline at the top of a small hillock. He was holding a strung bow in one hand and a dozen poisoned arrows in the other. When he saw who it was, he lowered the bow and beckoned us to follow him.

Just over the other side of the hillock we found Aman and Bai Taklabangan in a small, raised, open hut not much larger than a dining table.

'*Anaileuita*,' he said.

'*Anaileuita aleh*,' we replied, grateful for the chance to sit in shade.

'I am sorry I did not answer your calls,' Aman Taklabangan began, 'but I had to be sure it was really you.' He got up and unstrung his bow and put it, with the arrows, in between the palm-leaf roof pieces. He sat down again and moved his *parang* to his side.

'What's all this fuss?' I asked. 'Why does the place look so deserted? Why are people so scared all of a sudden?'

'Have you not heard the news?'

'You mean about the Pekai?'

'Yes, about the Pekai.'

'Well, all we know is what Aman Gogolay came and told us yesterday morning. There seems to be a band of people in the forest who are cutting people's heads off. It's ridiculous!'

'But it is true. Look, I have not actually seen the Pekai myself, but some people have, so I am simply telling you what I have heard.'

'Go on,' we urged. Bai Taklabangan, never a great conversationalist, picked up some four-inch periwinkle shells which she'd gathered at the coast and started cracking them open with her *parang*. She sniffed heavily and spat between the floorboards, then got up and disappeared into the tall ginger-lily plants near the hut.

Aman Taklabangan began. 'A few weeks ago a big ship arrived in the north at the harbour of Sikabaluan with supplies for the shop. On board were thirty foreigners who wanted to get ashore but who were not allowed to because they did not have written permission from the Governor. The ship was anchored a short distance out to sea but one night, when there was no moon, the foreigners leapt from the ship and swam ashore. No one saw them and by the time they were missed, they had disappeared into the jungle. Nothing more was seen of them.

'Nothing, that is, until two brothers were out collecting rattan to sell to Ali, the teacher at the coast. They had taken their bows and arrows with them and had managed to shoot a *simakobu* monkey. They separated briefly when one of them went to collect *obbuk* bamboo to cook the monkey in, but when he arrived back at their temporary camp he found that his brother had been killed. The head had been chopped off and taken away.

'Well, the brother rushed back to the coast and told the policeman. A search party was raised and they all went out to look for the murderers. After two days they found some strangers in the head-waters who were scared of being followed and in a fight they killed one of the villagers. His friends fought back and managed to kill five of the Pekai before the rest ran off into the depths of the forest. The villagers carried all six bodies back to the coast, buried the villager and gave the other bodies to the police. They told the whole village that if anyone saw a stranger, they should call to him and tell him to report to the police station because he did not have the right papers. If the man refused or if he did not answer, they should shoot to kill.'

'Good grief,' Alan exclaimed, 'the whole thing's amazing!'

Bai Taklabangan looked unconcerned as she climbed back onto the floor of the shelter. She was carrying a couple of ginger-lily fruit heads as large as clubs. She sat down, resting against one of the corner posts and started to extract the tart, juicy pulp-covered seeds from the fruit.

'You were hiding from us just now when we called; does that mean the Pekai have reached this river basin?' I asked.

'Oh yes,' Aman Taklabangan assured me. 'Children who have been playing in the river have seen men wearing long green coats moving around in the long grass. The village chief has seen them as well and found boot-prints in muddy places.'

'So they wear boots,' I checked.

'Yes, just like yours. They are green and leave a print in the mud. Their green coats seem to be just like the ones you sometimes use at your camp and in the forest.' He started to roll another cigarette.

'If they don't come to villages, what do they eat?' I enquired. 'Do they hunt or do they carry tins?'

'Neither. They do not really eat at all. They simply swallow pills, nothing else. The pills are like the ones we have seen you eat at your camp,' he stated, referring to our anti-malaria pills.

'What you're saying is that someone could mistake us for Pekai,' I said.

'Well, yes; but not always,' Aman Taklabangan replied, looking somewhat confused.

'What do you mean?' asked Alan in an effort to clarify the apparent possibility of getting a poisoned arrow through us from the hands of an over-zealous or over-nervous hunter.

'Well, they are not always wearing green plastic coats and boots,' Aman Taklabangan began; 'sometimes they wear long black wigs to make themselves look like women.'

'How stupid!' I laughed. 'I suppose they wear little dresses to complete the disguise!'

'That is so.' Aman Taklabangan said slowly with a questioning look in his eyes wondering how I knew. 'They dress up and when they think they are near people they get down on their knees and cry, and cry, and cry. When someone hears the cries they go to comfort the woman. The Pekai waits and if the helper is a man the Pekai will catch him, remove the long, hooked *parang* he has been hiding up his dress and cut the man's throat; slowly.'

'And what if the helper is a woman?' I asked, thinking of Jane.

'When a woman comes to help, the Pekai stops crying and catches her. He ties her up and gouges out her eyes; one by one.'

'So women aren't actually killed, is that right?' I tried to confirm.

'Yes, that is so. But I think I would rather die than not be able to see.'

The decision of whether to start our surveys was not easy. We had to begin them soon and yet there was something weird happening on Siberut. We were unable to believe anything Aman Taklabangan had told us; it was unlikely that even he believed everything he said. On the other hand the stories were likely to be founded on some truth.

By this time, Aman Taklabangan had almost certainly guessed we were going to ask him to be our guide but with traditional good

56

manners no one mentioned the possibility. Instead, we talked about the yield of his weed-ridden paddy field in which Bai Taklabangan was now picking the heavy seed heads. We ate some ginger-lily fruit and whelks, and listened to countless, unseen frogs singing from the hot, moist ground.

When the statutory half hour had passed I sat up and asked, '*Aleh*, would you be prepared to act as our guide on the first survey?'

He looked at the roof, the floor, the paddy field and then at us in contemplation. He wore an 'of course I'm going to say yes but I don't want you to think I'm easy so sit and suffer for another minute' expression, and drew heavily on his cigarette.

'Yes,' he said at last.

'Oh good,' we exclaimed in mock surprise.

'But,' he continued, 'I will want to go with someone else and I think it would be better if we travelled in this river basin where people know me and might know you.'

'That suits us. So, let's agree to leave a week tomorrow.'

'*Kawat*, good,' Aman Taklabangan agreed. We chatted further about local gossip and rice and then gave him a parting gift of tobacco. It was not much cooler as we walked away from the shelter, glancing right and left for long-haired wailing women with *parangs* up their skirts, and we were grateful to reach the cool shade of the forest.

Telling Jane the strange news was far from easy but, scoff as she might, she too realized that there must be a grain of truth to the stories, and was soon rummaging through our files to find letters of authorization to show any challenging hunters. During the following week we had to pack, copy maps and generally prepare both ourselves and camp for the temporary departure. Most important of all Sam and his family had to be found as often as possible so that they had no excuse for forgetting me.

We had many local visitors over the next few days who tried to persuade us not to go travelling until after the Pekai had been caught. But leave we did, to meet Aman Taklabangan in Sirisurak. Our locally made rucksacks were heavy and Stumpy was miaowing plaintively in a chicken basket tied to Alan's pack, for we had decided to leave him in the village during our absence. The paths along the ridges of Teitei Sabeuleleu and Teitei Jotjot were firm and dry, and a pleasure to walk along, but the final half hour of our walk was through the knee-high wet grass of the hill *ladang* or gardens which in many places were as slippery as the mud around camp. Jane and I carried on to find some shade while Alan stopped to photograph some of the garden crops. The people of the Mentawai Islands are possibly the only group in Asia who, after they clear an area to plant crops, do not burn the dying vegetation. Instead they leave it to rot down naturally, planting seedlings as soon as possible. By refraining from burning, they give the fragile soil a chance to consolidate itself under the slowly decomposing matted cover of vegetation. If the felled vegetation were burned, the soil would immediately be open to the elements and to the threat of erosion, and the gardens would have a far lower production potential. I'm not sure whether this unusual, but wise, practice is due to a greater understanding of the workings of the ecosystem, or simply because almost nothing there ever gets a chance to dry out.

We waited for Alan a few hundred metres from the village in the shade of a coconut tree, and while taking off our packs we heard voices of people approaching us and could make out the words 'Pekai' and '*teelay*'. Just then a husband and wife, whom we recognized vaguely, rounded the corner looking at the ground and pointing to our boot-prints.

'*Teelay*,' the man exclaimed when he saw us, and stopped in his tracks.

'*Anaileuita*,' we both said, smiling broadly.

'Er — *anaileuita aleh*,' he replied haltingly, looking at his wife for inspiration, trying to decide what to do next. There was an uneasy silence and then the wife put her hand into the large and obviously very heavy *opa* her husband was carrying and produced a nearly ripe papaya.

'*Maobak kaap pukop*, do you want to eat?' she said, offering us the fruit.

'*Tak leu*, we've only just eaten,' we lied with good manners. Then it suddenly struck us that this was a test, as the anxious expressions on their faces showed, because Pekai don't eat. '*Tapoi malaje'at kai*, we're still hungry,' Jane continued and she reached for the fruit. We made a great show of eating the crispy pulp and the couple looked more at ease.

'Where is your *parang*?' he asked.

'Alan's using it up in the *ladang*,' Jane replied, 'he'll be down soon.'

'A good one is it?' he continued, 'long, sharp . . .'

'No, not really,' I assured him. 'It's short and not really very sharp. Not like yours.'

'So it is not hooked or anything. That is good.'

Alan joined us and they were able to see for themselves the state of our *parang*. They seemed happier and we all walked into the village together. Aman Bulit and Aman Taklabangan were there to welcome us and we accepted offers of home-grown coffee. Over the drink the dominant topic of conversation was the Pekai. We learned that they didn't built any type of house or shelter but simply dug a shallow hollow and lay in that to sleep. There was disagreement on whether they merely cut the throats of their victims or whether they also took away the heads. Of those with the latter view, most agreed that the heads were placed beneath bridge supports to slow down the sinking of timbers.

The village chief came to see us and his status had obviously risen considerably since he had reported seeing three Pekai with his own eyes. The three village chiefs in our river basin had all reported seeing Pekai.

It was Aman Bulit who, quite out of the blue, added an important piece to the jigsaw of the Pekai that was forming in our minds. It was that Pekai is the way of saying PKI, the abbreviation for Partai Komunis Indonesia or Indonesian Communist Party and the inevitable tag for any group of insurgents. Suddenly the story increased its credibility. Although the Communist Party is outlawed in Indonesia, it was not inconceivable that pockets of resistance still existed. The dress and at least a few of the supposed habits of the Pekai which had been reported were not entirely incompatible with our view of what forest guerrillas would wear and do. Exactly why they should be on Siberut, however, was not clear.

After about an hour Aman Taklabangan introduced us to our second guide, Aman Jairaebbuk, and then the five of us set off, leaving Stumpy chasing chickens and getting walloped for his pains by their owners. Aman Jairaebbuk left the village wearing shorts and a shirt but after a few hours, when he realized we weren't going to disapprove, he changed into a loincloth, thus showing off his fine tattoos. Both he and Aman Taklabangan carried a bamboo quiver decorated with tufts of monkey fur and a six-foot bow of polished *pola* palmwood slung horizontally from their left shoulders. As we walked due south from Sirisurak along the banks of the Sirimuri river they collected leaves and bark from *ipoh* bushes to renew their arrow poison.

The reserve we were going to propose for Siberut would be designated primarily for the rare endemic wildlife, but we had decided that trying to count animals in different areas would be impractical in the few months available. Instead, we worked on the assumption that a continuous tract of relatively undisturbed forest would contain all the species we were concerned with. It was therefore important to ascertain where and how the people lived. How far did they venture into the forest for different activities? How disruptive were these activities? What was the traditional system of land ownership? How far up each river was it necessary to travel before a continuous canopy formed above it, enabling the primates to cross? What evidence could be found of the effects of hunting on the animals?

We had decided that on this and the other surveys we intended to make on Siberut, we would stop every half hour to make notes. This was much more practical than trying to estimate distance so that we could stop, say, every mile. At every stop Jane sampled all the trees within twenty yards for signs of feeding by *jirit*; Alan noted the general form of the forest or other vegetation, listing predominant species and the extent of human disturbance; I was left to make sure that I could plot our position accurately on the map, and to note any mammal signs such as tracks, faecal smells, half-eaten fruit, or calls.

By late afternoon we'd reached an empty *sapo* in the headwaters of the Sirimuri and decided to stay the night there. Aman Taklabangan searched through the perforated roofpieces of the thatch and eventually discovered an old wooden platter on which he piled the *ipoh* leaves and bark. He chopped these finely with the well-honed tip of his *parang*, and then Aman Jairaebbuk added some rough-cut pieces of small red chilli peppers and derris tubers he'd collected in the overgrown clearing. Searching through the thatch he found a heavy wooden hammer which Aman Taklabangan used to beat and grind the ingredients together.

After a good five minutes of hammering, Aman Taklabangan said, '*Kawat*, that is well mixed. Now, Alan, you must pass me those two wooden spatulas hanging near the hearth. We use those to squeeze out the poison.' Aman Jairaebbuk took one of the foot-long spatulas from Alan and on the broader end rested a turk's head ring knotted from a length of shaved rattan. He transferred the poison mixture into the centre of the ring and laid the second spatula on top of this. Next he bound the ends of the spatulas nearer the ring loosely together and used the other two ends like the handles of a large nutcracker. As he pressed more and more firmly, so a light reddish-brown fluid oozed through the ring and was collected in a small bamboo container.

When they were certain that every last drop had been extracted,

Aman Taklabangan lit a smoky fire. Together they then took some of the unfletched arrows out of their quivers and proceeded to coat the tips with poison. When each was covered, it was laid above the gentle fire and when the fluid had dried to the colour of dark mahogany, further coatings were applied.

While they were doing this I noticed that there were several types of arrow tips and so I asked Aman Taklabangan whether that had any significance. 'Oh yes,' he said in mild surprise at my ignorance, 'we use different ones for different animals. The ones left in the quiver, they are for squirrels; those with small barbs are for the *soksak* so that the arrow is not pulled out as it runs away in the undergrowth, but for *loga* only this plain tip is necessary because when it falls out of the tree already it is dead. That arrow with a thick, blunt tip we use for birds. Now those heavy arrows with metal tips drying over the fire, they are used for deer and wild pig. The other poisoned arrows with slightly barbed *ariribuk* tips used to be used for monkeys because when they tried to pull them out the shaft breaks off from the poisoned tip. Then they die quickly. Hunting monkeys is now illegal because people like Mr Tilson think there are too few monkeys. Anyway, the best way to hunt monkeys is to use dogs that either catch them when they run on the ground or keep them up in the trees where they are easier to shoot. But there is not a single good hunting dog left in Saibi.'

'So if you no longer hunt monkeys,' said Alan, 'how come you've just poisoned those barbed arrows?'

'They are for the Pekai,' Aman Taklabangan replied in a matter-of-fact tone.

When we retired for the night our guides positioned their mosquito nets on either side of ours, keeping their strung bows within easy reach just in case of any marauding Pekai (or, perhaps, a succulent deer). Neither disturbed our sleep.

The next day could hardly have been wetter but we decided to slip and slide our way on to the next major river to the west. Walking over Siberut bridges, usually no more than felled saplings, was tricky enough in dry weather, but when rain was pouring down, rivers in flood were washing over them, and our feet were covered in mud, there was little hope of staying upright. We were soon so utterly soaked that we simply followed the pig paths – down one stream bank and up the other side, much to the amusement of our guides. They, of course, strode over the saplings as though they were as stable as Westminster Bridge, but were very understanding. Late in the afternoon of that miserable day we met two sisters who were setting traps at the mouth of a small flooded stream. These would catch the fish as they returned to the major river when the water receded. The women offered us hospitality and a roof for the night; with no hesitation we gratefully accepted and followed them upstream. They walked incredibly quickly and seemed to skip over the deep mud along the banks of the river. It must have been something to do with their splayed feet and the fact that they weren't carrying packs, because Alan, Jane and I kept sinking in up to our knees and having to enlist help from the others to pull us out; even our guides had some trouble. Dusk was falling when we eventually saw the large *sapo* that we were aiming for, perched high on a bank at the apex of a large meander of the Sakerake river. At last we were able to change into dry, thick *sarongs* and good humour abounded.

The house owner was Teteu ('Grandfather') Songai, one of the oldest men in our river basin, and although *compus mentis* and still able to get around, he was thoroughly looked after by his equally aged wife and his two widowed daughters. We offered around tobacco and sat on the veranda, with a fading panoramic view of the swollen, brown river below, and the dense sago swamps along its banks. Teteu Songai was wearing a beaded headband, the 'insignia' of a *sikerei*, and a beautiful set of red beads around his neck. His body was fully tattooed; the spreading and blurring of the lines attested to his age, but it was difficult to put a figure to it. He was still the major pig and chicken owner in the area, and old though he was, he was still useful to his family and clan. He spent his days feeding the animals, talking to them, chasing off other people's pigs if they came to eat his stock's sago and generally cosseting them.

Teteu Songai spoke softly to Aman Jairaebbuk and the latter went out briefly with a length of thin rattan cane. All but the last foot or so of the cane was quite rigid, the tip having been pared and tied into a noose. We watched him approach a large white cockerel that was roosting in a low fruit tree growing next to the house. He tickled the cockerel's belly with the noose, causing it to lift its leg and allowing him to fix the noose loosely around its thigh. With a flick of his wrist, he brought the bird to the ground and caught it. When he had brought the splendid white bird to the hut, Teteu Songai asked me if I would kill it. I hesitated at this unexpected invitation that I suppose was the equivalent of being asked to lay the table, and I passed the buck as quickly, but as politely, as I could to Aman Jairaebbuk. I was exceedingly glad that I had because he killed it so gently, almost lovingly. He nestled it in his arms, holding and stroking the neck with his hand, and spoke softly into its ear. He told it not to forget how it had been cared for since it hatched and how it had been fed regularly on grated sago and coconut. It had been allowed in the house to scavenge for scraps and the only reason it had to die was because we were hungry. Then Aman Jairaebbuk brought the neck forward and down over his hand and within a short time the bird was dead.

Jane was doing her best to help the women to wrap sago flour in leaf parcels prior to baking them on an inclined rack over the fire. The women were very understanding of her gauche efforts and gave Jane parcels that were nearly completed to finish off. As she got the hang of it, so she was given increasingly incomplete parcels. Very occasionally one of her parcels came undone and once, in desperation, she uttered a barely audible *'Teelay'*. The house went very quiet and still, and then the women started laughing. *'Teelay'* is not so much a swear word as a common expletive the approximate meaning of which is 'the most private part of the female anatomy'. It is used so often, in fact, that it was the first word that anyone ever heard Aman Taklabangan's son say. It is not particularly polite, however, to use it when visiting people in their homes but the obvious frustration that Jane was experiencing when wrapping parcels of sago seemed to be excuse enough for its use. Jane turned bright red when she realized what she had said but then she, too, started laughing.

Some time later we were presented with supper. We all sat cross-legged around two long wooden platters and a pile of sago parcels. The platters had stewed chicken at one end and boiled roots coated in grated coconut at the other.

'Kawat, mukopita, let's eat,' Aman Taklabangan said as he launched himself towards the sago parcels. With a stick of sago safely in his mouth, he handed out two or three parcels to each of us. The leaf around each of the parcels was charred and brittle. We watched the others rub their hands along the parcels to remove the leaf bindings and imitated them. Jane was the first to discover that it wasn't quite as simple as it looked. There are curved spines along the leaf-rib and if you inadvertently rub the wrong way you cut your hand. Everybody smiled goodnaturedly and showed us how to do it correctly. Even so, in the smoky gloom of the house, I don't think any of us remained

totally unscathed. The baked sago itself was delicious – hot and soft inside but crispy on the outside. Perhaps it was that we'd been on Siberut for too long, but I could close my eyes while eating hot sago and believe I was eating hot buttered Hovis.

We took pot luck when picking up bits of meat and more than once we ate portions we couldn't identify; only the contents of the intestine, the gall bladder and the trachea had been excluded from the stew. I watched Alan pick up a chicken's foot and look aghast at what faced him. It seemed to be all he could manage to lick the gravy off the claws. He bit gently into the scales but could find nothing obvious to eat. Trying to look nonchalant, he slipped it into the palm of his hand and picked up some more sago. Then, hoping no one was looking, he leant back and flicked the foot deftly towards an eagerly awaiting dog. No one noticed.

Aware now of the dangers of picking blindly, I made sure that I felt a soft rounded piece of meat before I took it off the platter. I got the head. I returned its steady gaze for a while and pulled bits off the neck. I couldn't make myself eat the comb or put the whole thing in my mouth and feel the open beak against my teeth. I slipped it to the pigs below the floor.

Aman Taklabangan put away far more than anyone else in half the time. We finished after everyone else because our jaws were so unused to chewing anything. Our diet in camp was largely tinned meat and rice which required no more than forking off the plate and swallowing, and consequently our jaws ached with the effort of eating real meat and sago sticks.

'*Kawat, asageat ka'aku*, I've had enough,' Aman Jairaebbuk declared loudly.

'*Ka kai leu*,' I agreed.

One of Teteu Songai's daughters cleared away the platters and pushed the sago wrappings and other dross between the floorboards. The dogs were favoured when one of the old sago parcels was unwrapped for them. They leapt on it amid great yelping and one grabbed the lion's share then rushed out to hide in the bushes. Meanwhile, Aman Jairaebbuk, Aman Taklabangan and Teteu Songai started to talk about the Pekai again and swapped stories of horror. They were discussing the means by which the Pekai attracted hunters so that they were able to cut off their heads.

'I have heard someone say,' Aman Jairaebbuk began, 'that they can imitate animal calls.'

Joja

'*Teelay*,' Teteu Songai croaked, 'but you can always tell the sound of a man from the sound of an animal.'

'Not with the Pekai, Teteu,' Aman Jairaebbuk corrected. 'One of them, so people say, climbs a tree and before dawn he cries out "*bagok bagok bagok bagok gulugulugulugulu*" just like a *joja* monkey.'

'*Teelay*!' the other two chorused, 'then what?'

'Well, it is such a good imitation that we would try to creep up to the *joja* hoping to shoot it. At dawn he calls again, "*bagok bagok bagok bagok gulugulugulugulu*" and the hunter gets closer and closer, hoping to get a clear view when the *joja* leaves the sleeping tree.'

'Why doesn't the hunter shoot the Pekai though?' asked Teteu Songai.

'Simply because the Pekai on the ground wear green mantles and walk quietly. And then they catch the hunter, and *amatei'at ia*, he's dead, very dead.'

'Do they eat the hunter?' enquired Alan.

'No,' explained Aman Taklabangan, 'they just cut off the head and take it away.'

Up to the first quarter of this century, Siberut men were doing the very same thing on their own head-hunting raids. Inhabitants of neighbouring river basins, and neighbouring islands, always had running feuds between different sets of clans and the normal way to work off rising feelings was to make a raid on an enemy group. The preparatory ceremonies were elaborate and if members of nearby clans had similar grievances with the same people they too would send representatives to the organizing clan. Specific taboos were observed and after a week or more of ceremonies the final sacrificial animal was slaughtered. If the *sikereis* found good omens in the blood vessels of the heart and gut mesentery, then the force would leave for the enemy territory.

They carried an array of weapons and wore flowers and leaves both for decoration and for spiritual protection. Unlike many other peoples from whom head-hunting has been reported, little store was set by whose head was taken and so the first man, woman or child of the right clan who was seen was killed, usually with a poisoned arrow. In the event of close combat, when arrows became impractical, the long razor-sharp *parang* was employed. If that was knocked from their hands, they would resort to a *parittei* or dagger. If these failed them, it was either too late or else they ran away. From what we were able to gather, heroism has never been a favoured attribute of Siberut people. The older men would take heads leaving the younger men to claim the hands and feet as trophies. In the event of a raid being unsuccessful, they would apparently rid themselves of pique by shooting at coconuts instead!

The return of a successful raiding party was greeted by great rejoicing and further ceremonies were organized; this time with the heads and other body parts placed prominently in the *uma* to 'watch' the dancing and singing. After two or three days the heads and other trophies were taken a long way away and buried. Siberut people never seem to have been cannibalistic.

It wasn't just humans who slept in the shelter of the house that night. Forty-two pigs of every size and several colours slept about three feet away from our ears and although the ridges of the floor were somewhat smoothed by sleeping mats of flattened sago bark, we stayed awake for a long time. By peering through the gaps in the floor you could see the pigs piled in a great heap. I suppose that to have pigs beneath your house is similar to keeping money in your mattress — except for the noise. Pigs are not penned on Siberut but are fed regularly in the enclosures beneath the *sapos* to keep them nearby and tolerably tame; only the full-grown boars roam the hills. One of the pigs beneath Jane and me was a huge castrated porker who was presumably just biding his time before the next important feast.

The next day we came across a number of small field houses and our guides told us it was quite acceptable to go inside and take a look around although the owners weren't home. In every house there were monkey skulls hanging from the roof, and we counted the number of each species present, telling them apart by differences in the eyebrow ridge and jaw length. We found only one *beelow* skull during the whole of this preliminary survey and at one of our periodic rests to eat a stale stick of sago I asked Aman Taklabangan why this should be.

'People have never killed many *beelow*,' he began, 'but I am not sure why. The songs they sing before dawn sound like ghosts, and nobody likes to go out then.' I had to agree. 'It also looks very like a human. Anyway, it is taboo to eat *beelow* in an *uma* or *sapo* and the one skull you saw earlier was presumably killed by someone who has rejected the traditional beliefs. Mind you, I think you'd find that most men have eaten *beelow* at some time in their lives. Young people are not so affected by the taboo, and if you cook and eat it in the forest, who is to know?'

The most westerly inhabitants of the Saibi basin live at Simabuggei, 'the sandy place'. One of three families is headed by Mr Tseng, a Chinese shopkeeper, the last person we expected to meet in such a remote spot. His house was very comfortable and he invited us to sleep on his blissfully flat floor of sawn planks that night.

Who, you might reasonably ask, needs a shop in the middle of Siberut? The question was answered for us when a line of hunched figures appeared on the brow of the nearest hill, silhouetted against the orange evening sky. They were all carrying large packs on their backs and a few minutes later the sleepy shop had become a hive of noisy activity and eager business.

The people had come from Simatalu, the river basin due west of Saibi, over the watershed that runs approximately down the middle of Siberut. They set down their loads; hemispherical, brown, sweet-smelling dried coconut meat or *copra*, bound tightly together inside plaited palm leaves fitted with straps of thin bark. The mens' packs weighed eighty pounds and the womens' only a little less, yet we heard not a single groan that could have been interpreted as, 'good grief, thank heavens I've got that off my back'.

The formation of the Simatalu river mouth is such that it can't be negotiated by the eight-ton boats that enter all the other harbours on Siberut in at least some months each year. Thus, the only sure way the

67

basin's inhabitants can make any money to buy the tobacco, cloth, pots, lamps, axeheads and rice that Mr Tseng offers, is to sell him *copra* for making cooking oil.

We greeted our fellow travellers as they came onto the shop's veranda, but from their first couple of sentences it was clear that communication was going to be limited to smiling and pointing. They spoke what seemed to be a totally new language with only a few words that we could recognize. We later learnt that there are four languages on Siberut and twelve distinct dialects. That such differences should have come about illustrates how little contact people from different areas had and, to some extent, still have. Even Aman Taklabangan and Aman Jairaebbuk were having difficulty communicating.

The women, who ranged in age from about sixteen to twenty-five, wore only cloth skirts and many were tattooed from neck to ankle in thin blue lines. The front of their shoulders formed a focus for up to ten lines, some of which curled up to their jaws; some fell across their chests like meagre mayoral decorations; and others ran part way across their backs. A long straight tattoo extended from below their lower lip down their necks to their navels, and another crossed this horizontally between the nipples. The back of their hands and wrists bore patterns like embroidered gloves and their calves had dense bands of patterned blue in anklets around them. Between and beside the lines, crosses and star shapes were common, but apparently subject more to individual whim than the conventional designs each of which had a specific name.

The men wore red-dyed, bark loincloths that ran between the legs and buttocks, around the waist, and finished in a flap in the front to hide any indecent shapes. The mens' tattoos were broadly similar to the womens' but their chest lines were more in the form of a breast plate and their buttocks and thighs bore numerous parallel, horizontal lines, rather like the instructions in a pictorial guide to carving, that accentuated the contours of their taut, muscular bodies. The majority of the visitors had also had their incisor teeth filed to V-shaped points so that when they smiled, which was often, they looked like latter-day draculas.

Tattooing is one of the practices that has been frowned upon since the first governmental contact in the early part of this century. In those river basins that are relatively remote, such as Simatalu, such views have been slow to percolate, whereas in Saibi there are few men under thirty with tattoos. Thus, Aman Taklabangan and Aman Bulit have very few lines but the older Aman Jairaebbuk and Aman Gogolay are fully tattooed.

They chatted far into the night, chain smoking around a dim lamp. Both men and women kept the palm leaves they used as cigarette papers in holes through their earlobes and these steadily emptied. Lying on our sleeping mats we heard Aman Taklabangan telling the assembly about all the crazy and inept things we did: like falling off bridges, nearly poisoning ourselves with *lutti* root by using the water it was boiled in (nobody told us not to, because surely *everybody* must know), steering a dugout in circles and naming our cat and feeding it off a small plate; all of which were greeted with peals of laughter. He also told them in a grandiose manner all about the Pekai. These people hadn't yet heard the news but you can bet that within a couple of days there wouldn't have been a single person in Simatalu to whom the stories hadn't been passed on.

Six months later Jane and I were paddling up from the coast to Sirisurak with supplies in a flooding river. Even hugging the inside banks it was very heavy going with such an over-laden dugout despite the efforts of our three helpers. So we were excited and relieved to hear the throb of speedboat engines coming up behind us. Speedboats are seldom seen on the river and there is always much speculation regarding who it is, why they are coming, what they are bringing, and so forth. Our major concern at that time was whether they would be willing and able to take us and our stores on board. A quarter of an hour later we watched the speedboat round the corner behind us and were amazed by what we saw; it was a broad, new fibreglass boat, full of fourteen dark-clothed men wearing crash helmets with smoked-glass visors – like the river watch of a fictional paramilitary organization. The boat was using two huge 40hp engines which were only appropriate because of the flooded state of the river. As the boat slowed down so that we could talk, we recognized a number of local government officials we knew from Muarasiberut. As we had hoped, they offered us a lift and our hired paddlers were glad to see us go.

My conversation with the most senior among them was none too easy, what with the roaring engines, splashing and surging water and orders being shouted to the helm to avoid the half-submerged logs being washed downstream. His crash helmet and visor didn't help communication either, and they would have been more of a liability than an aid had he actually fallen in the river. I nevertheless managed to catch up with local gossip, and then broached the subject of the Pekai.

'What really is the truth behind all the stories we've been hearing about the communist guerrillas, the Pekai?' I asked.

'Oh that, I shouldn't worry about that. Communists have never been on Siberut and probably never will. The only political party on the island is Golkar, the government party.'

'But then why are there so many awful stories about terrorists killing Siberut people and taking heads?'

'Pardon,' he yelled as the engines whined along a straight stretch of river.

I put my mouth next to the holes beneath which I assumed were his ears and shouted, 'I said, how come so many people think the PKI has been on Siberut for months killing people?'

'Oh that,' he replied with the slit of his visor in my ear, 'there never have been any PKI on Siberut. The stories all started, I believe, when some Filipinos were surveying for a timber company in the north of the island. They met a hunting party and both sides were so shocked that each ran away from the other as fast as possible. I think that is the truth.'

Maybe, but wherever we went on Siberut everyone spoke of the unbelievable evils of the Pekai, the hooked parangs, the stolen heads, the pills, their skill at imitating animals and their boot prints. Although some Siberut people have a type of plimsoll, very few if any have jungle boots with deep treads, and our surveys probably kept the stories fresh as our boot-prints were seen.

The Siberut people also tell stories about a huge black, hairy, cunning, vicious apelike man called *Silakokoinak*. He did evil things to people but no longer exists in body, although he does still exist as a ghost and on a still night one can hear him flying over the trees. I wonder how long it will be before the Pekai become ghosts.

One thing is for sure; now we know what people did before they invented television.

Chapter Five
Getting to Grips with Gibbons

I sat on a damp, leafy hummock and scraped around the bottom of the jam jar to get out every last grain of sweetened red rice. I had hoped that Sam would be singing early but he didn't seem even to be awake

Sam singing

yet. So there was nothing to do but eat breakfast. I was now able to keep up with the group for a whole day barring mishaps, and they had grown very tolerant of my stumbling around beneath them. Sam had entered the night tree in front of me at 4.30 the previous afternoon and from the vague silhouette that I could make out, he didn't seem to have moved a muscle in the twelve hours since. As usual, because Bess and Katy had been in front of Sam I wasn't able to see exactly where they were sleeping, but they always seemed to sleep nearby.

Just as it became light enough to scribble down a few notes I felt some drops of water on the back of my neck. I was in the process of writing 'light rain' when I was hit on the head by alternately hard and soft missiles; the air around me took on a rather fruity smell and my notebook was speckled with patches of brown. Now I knew that Bess and Katy had been sleeping directly above me. Similar dropping sounds came from below Sam, and I was clear that my *beelow* had all woken. You read of predators that roll in their quarry's faeces to disguise their own scent, but I felt that in the circumstances it was hardly necessary for me. I did what I could to remove the remains of their last night's meal of nutmeg fruit from my hair, but before I could finish, Sam piped quietly and leapt out of his tree. He was followed closely, and soon overtaken, by Bess and Katy.

They headed for a large spreading tree that bore a fruiting fig. The figs were quite small, little more than a fifth of an inch across and the colour of strawberry ice-cream. The *beelow* sat on and hung from the branches, never going back to any they had already visited that day. Apart from simply eating the figs, they also felt many of them, squeezing them gently between their thumb and the side of the index finger. It may be that they were testing them for ripeness but even I, from the ground below and with binoculars, could tell the ripe red figs from the unripe green ones. Gibbons can see colour too and many of the forest fruit change colour when ready for eating. So, if squeezing fruit is not necessary to test their ripeness, they might be breaking the 'don't squeeze me 'til I'm yours' rule. That is, they might be slightly bruising the unripe figs that are within their reach to ripen them faster. By so doing, more suitable figs would be available to them next time they visited the tree, probably the following morning, and their time in the tree would be better spent.

The normal sounds of the morning were interrupted briefly by the harsh repetitive barks of the *bokkoi* macaque. These monkeys move around in loose groups of thirty or so, but more often we saw splinter groups of eight to ten. The barks, which seemed to be getting closer, were given by the leading male of the group, probably to let all the sub-groups know his position. The *bokkoi* came to our study area every few weeks and their whole range must have covered about two square miles.

Behind me, Sam started to hoo softly and since he no longer hooed at me, it must have been because he could see the *bokkoi*. I followed the line of his gaze and saw some tree branches moving. Then seven *bokkoi* came into view walking loutishly and strongly. At first glance they were similar to dogs; their limbs were all about the same length, their muzzles slightly elongated and their tails quite short. The females were heavily built but the big males looked like muscle-bound boxers, the likes of whom you'd not want to meet in a dark alley. All seven of them descended on a small fig vine and fed on the fruit as though it were their first food for days. They grabbed the fruit greedily with both hands passing it to their mouths as fast as possible. Only some of it would be eaten immediately, however, the rest being tucked into cheek pouches that allowed a fruiting tree to be exploited faster than they could swallow and after their stomachs were full. In this way a *bokkoi* group could easily remove all ripe fruit from a tree in a single sitting. Later, in some less frenzied moment, they would use their shoulders to

Bess eating vine fruit

prise food out of their cheek pouches and chew it slowly. The major male barked again, further away this time, and the *bokkoi* I could see thumped off along the ground towards him. I discovered that evening that Jane had come across this group a couple of hours later on the Flat Trail. Creeping closer to get a better view and to count them she had been startled by a rustle in the undergrowth just to her left and froze as a large male walked nonchalantly past her, less than a yard away.

Sam and Bess watched the receding *bokkoi* with disinterested stares and then resumed feeding. Katy hadn't bothered to watch them and now that she was full, she sat quietly until her parents had finished. Although it had been cloudy, the sun was now making valiant efforts to make its presence felt; indeed, it must have been four days since we'd seen the sun and the last eight nights had been wet. Sam hadn't sung much if at all before dawn on any of those days, probably because he had felt too cold, wet and dejected.

Katy brachiated somewhat gauchely off towards the West Ridge path, snatching at the odd juicy insect grub as she went; she was growing fast and becoming increasingly independent of Bess during the day. When Bess and Sam followed her, the drongos – slim, crow-like birds – fluttered after them, occasionally diving off at an angle to catch a cicada or some other large insect disturbed by the swinging *beelow*. When one was successful, it found a perch and clubbed the insect on the branch as a kingfisher might deal with a fish. There was never a day when I didn't find drongos around the *beelow* but, unfortunately, the presence of *drongos* in an area didn't always mean there were *beelow* nearby.

Sam and his family were now in a medium-sized *latso* tree where they had first fed about two weeks previously. Since then I had found them feeding in no fewer than twenty trees of the same species, all within about two acres. Although the *beelow* doubtless knew their eighty-acre home range extremely well, they could not possibly predict when and where fruit would be available. There were at least two hundred and fifty species of tree in their area and although there were some times when more trees fruited than at others, there was no great seasonality in fruit production. The *latso* trees fruited more or less every year, many figs fruited every few months, whereas the tall dipterocarp trees on the high ridges probably fruited only once every ten years. In addition, individual trees of some species all fruit together, like *latso* and dipterocarps, but figs, for example, showed little or no synchrony. The pattern of availability of the fruit on which the *beelow* depended was, therefore, very complex, and it was only by ranging over all parts of their forest regularly that they were able to see what fruit was coming on to ripen and which areas would be worth monitoring for a few weeks. With nearly six thousand mature trees within their domain they would undoubtedly be fallible, but by watching where flocks of frugivorous hornbills and pigeons settled they would probably very rarely miss out altogether on a source of fruit.

Sam took hold of a light-green fruit, twisted it off the branch and bit off one end of the rind the inside of which was, for some reason, vivid pink. *Latso* fruit are about two inches long by an inch wide, and in cross-section look like a seven-pointed star. Removal of the rind reveals a translucent-white, juicy pulp and the *beelow* popped this in their mouths and dropped the bottom half of rind. They chewed little, swallowing it quickly, seed and all, and then moved on to the next fruit. Many fruit turn red to advertise their presence and ripeness, and it would seem that the green *latso* fruit is an exception. In fact, almost a third of the fruit eaten by *beelow* were green when tolerably ripe but almost all these were conspicuous in some way other than colour — *latso* fruit were a distinctive shape, very different from the leaves, others were very small and round but stuck on the ends of an erect structure looking like a miniature Christmas tree bedecked with baubles, and yet others were lumpy, warty, spiny, spotty, skinny or dumpy. In contrast, many of the reddish fruit were small, smooth and spherical.

Gibbons aren't the most dextrous of beasts, since their hands are primarily a biological hook designed for brachiation, and now and again the *beelow* would drop *latso* fruit after removing some of the rind. I dusted these fallen fruit and ate them gladly because they were the most tasty element of the *beelows*' diet; somewhere between a lychee and a sweet lemon. I tried the taste-test for most of the fruit my *beelow* accidentally threw to me and more often than not found them bland or downright unpleasant. We're hopelessly spoilt by the fruit industry that can supply us with juicy sweet apples, strawberries and oranges. For Sam and his family, fruit eating was always a compromise. Eat them too early and you might do yourself no good by taking in too many of the toxic chemicals which plants use to protect their 'seed containers' before they've matured. Wait too long and you find either that some other wretched fruit eater has beaten you to it, or that the fruit has begun to ferment. Life is never easy.

Having all had a good share of *latso* fruits from two or three trees, the *beelow* travelled directly to the tall trees on the West Ridge to rest. I sat down where I had at least one of them in view. Even at that stage I was only just able to tell Sam and Bess apart at a distance. They were both totally black, very little different in size and their genitals were small and surprisingly similar when the view was less than perfect. Oh, how I longed to be able to call all the *beelow* in the study area by appropriate names such as Baldy, Scarface, Patch, Stripey, Dotty, Bigfoot, One-eye and so forth, but they were more alike than black beans in a pod. Some other gibbon species have coat colours which differ not only between and within the sexes but also with age. As it was, I found that only by long association with the *beelow* did I learn the small differences in their faces, in the way they moved and in the way they reacted to situations.

Sam (or was it Bess?) was sprawled on his back on a large bough, one hand pillowing his head and the other dangling in space. He lay in unobscured sunlight and after a while steam began to rise from his damp fur. If I'd been able to see more clearly I'm sure a smile of quiet contentment would have been visible. Inland, Siberut is wetter than most of the forests where gibbons live, receiving as it does about fifteen feet of rain each year – fifteen feet is taller than a double-decker bus. Luckily it doesn't fall all at once, although sometimes it feels like it, but in four indistinct seasons: two wet, one very wet and one incredibly wet. In areas with less rain than Siberut, sunbathing by gibbons isn't very commonly seen.

Gibbons are terribly restless animals, never being able to concentrate on one activity for more than a quarter of an hour. Never, that is, except when resting in full sun and with a full stomach. Then rests can last several hours. Katy, typical youngster, was easily bored and I watched her playing games on her own; leaping at vines, twisting from one hand and even hanging from her feet. She leapt back to the tree, sneaking up onto Bess who was resting peacefully in the sun, and jumped onto her belly. Bess clouted her and opened her mouth slightly in a threatening gesture but this hardly deterred her daughter. Katy tried again and this time received a rather more positive response; being grabbed and rolled onto her back at which she began laughing in a 'stop it, stop it, I love it' way. Bess soon became bored with this and Katy went off to play on her own again. Unlike other apes and almost all monkeys, young gibbons lack relations of the same age to play with. This is a result of the monogamous mating system of gibbons which, unlike the human monogamous system, is combined with strict territoriality precluding the meeting of youngsters. Many gibbon groups have two offspring but their ages will almost invariably differ by about two years. So, Katy had to play by herself or with Bess. The reason she didn't play with Sam was illustrated at their next feeding tree when Katy and Sam found themselves within touching distance of each other. Katy got short shrift from him; he bared his teeth producing a weird, short, ghostly howl, and Katy rushed off screaming to Bess. In many ways, Sam led a very lonely existence. Unlike other primates, *beelow* seem to groom each other very little (though they often groom themselves) and I only saw Bess and Sam embrace once. That was early one miserable morning after Sam had moved from his sleeping tree to hers. Sam embraced Bess who was cuddling Katy, but it was probably more to do with getting warm than with affection.

At times now, Sam was lagging about fifty yards behind Bess and Katy and I had to rush back and forth to check up on what each was doing. Sam was occupying himself in a *payleggut* tree whose flaking bark made the trunk look as if it was swathed in layers of thin red tissue paper. He studiously tore pieces of the bark from the trunk, watching carefully to check if anything edible had secreted itself beneath, and then dropped them like a child with a sweet-wrapper. Sam was probably eating insect larvae but wouldn't have been above trying gecko eggs, small lizards, spiders or insect galls. He had a

Sam on *payleggut*

surprisingly low success rate so whatever he got from it must have been worth his time and trouble.

By now, Bess and Katy were up at the Glade picking at dead leaves and loose bark themselves. From the tall trees in which they were sitting they could see for miles to the south and west and from over the Paitan I could just hear the first notes of a female song. Sam arrived in their tree and all three of them looked out across the valley. A female to the north-west, probably on our side of the Paitan, started to call and soon I could hear about four females – I assume Bess could hear them too.

Then Bess made noises like a girl learning to yodel (unsuccessfully). She floundered around for nearly a minute and then broke into her first introductory notes. Each lasted rather more than a second and was even pitched. She produced about forty of these notes each minute for several minutes climbing higher still in the tree as she did so. Then, like a car being eased into third gear from a whining second, she flowed smoothly from the introductory notes into her full song. An eminent 'gibbonologist' has described the song of the female *beelow* as 'the finest music uttered by any land mammal'. Even allowing for my obvious and unashamed bias, I am bound to agree.

The pitch of the notes rose slowly over about ten seconds, getting shorter and developing from simple hoos into whoops. Then the pitch stopped rising but the notes continued to quicken and she leapt around the tree, swinging wildly and tearing leaves from the branches. At the climax she was singing five loud, high notes every second and leaves and debris rained down onto me. After about ten seconds of the trill she slowed down and the notes got longer, lower and quieter.

Bess then sat down to rest but less than half a minute later she started again, this time without any introductory notes. This 'great-call' lasted even longer than the first, and Katy joined in the acrobatic display. Katy accompanied Bess during the third great-call, never quite getting the same note as Bess and never quite managing to produce whoops as fast as her at the climax but she showed great promise. Between great-calls Bess gave single whoops and sailed up to the notes of the great-call from these. Thus she continued for twenty minutes or so; the length of each great-call remained more or less the same but the gaps between them became longer and longer, presumably as she tired and became less inclined to rush around.

A single distant female could still be heard but all the others had stopped. It's tempting to suggest that females sing together to form an

organized chorus. However, although on average Bess and her neighbours sang every fourth or fifth day, I'd known days when Bess sang
but all the other females in earshot stayed silent and, conversely, days
when every female *beelow* on Siberut seemed to be singing fit to bust
but Bess carried on looking for insect grubs. It's likely that a female
has to be really well motivated to sing so that if the weather is not too
bad and she hears another female sing, then maybe she'll be moved to
sing. On the other hand, if she had called the day before, even bright
sun, a full belly, dozens of singing females and time to spare might not
be sufficient incentive to make her open her mouth.

Bess was taking a well-earned rest in a nearby tree and Sam was
starting to give short piping notes, each separated by almost a minute
and he was looking for grubs between times. A few other males within
earshot were also starting to sing and Sam progressed through the
elementary stages of his song quite quickly. An hour later he was still
going strong and didn't have time between the 'lines' of his song to
feed. Katy was a bit restless but Bess sat still. Sam was still too, in great
contrast to the great-call display, although he did turn his head round
so that males all around could hear.

Ninety minutes after starting Sam would have done better to have
stopped. He was developing a frog in his throat and coughed in the
middle of phrases, sometimes picking up where he had left off, sometimes giving up and starting again. The complex trill seemed to irritate
his epiglottis and I felt quite ashamed to be associated with him.
Eventually, he bowed out gracefully and brachiated over to sit with
Bess.

Whereas Bess sang every four days on average, Sam sang every two
days, more often than not before dawn. *Beelow* are unique among
gibbons, unique indeed among all monogamous primates and most
other animals that live in permanent pairs, in that the adults do not
sing duets. Instead they sing quite separately; the fact that Sam sang
after Bess on this occasion was probably little more than chance. Duets
are supposed to cement the pair bond, like buying a wife flowers on
Friday, but *beelow* appear to live quite happily without such frip-
peries. Exactly why this should be so remains a mystery.

But why should they be singing at all? Is it just joie de vivre or is
there a message? Well, they may well enjoy singing, but the real
purpose behind the songs is defence; the male directing his song only
towards males and the female directing her song only towards females.
The home range belongs to Sam and only to him. He's the one who

checks up on the boundaries, who is expected to chase out intruders and who deters itinerant *beelow* from approaching. He is saying 'keep out' but that's not all. He could advertise his presence in a fraction of the time he actually sings for, yet he goes on and on making his song more and more complex. It doesn't seem to make sense.

In many animals, territories and mates are fought over and the size of the combatants is an important factor in who wins. In many cases disputes are settled without a blow simply by the adversaries weighing up their respective chances of winning by judging the size of the opponent. Large size can therefore be an advantage. For gibbons, large size would be a distinct disadvantage because they depend on their agility and light weight to exploit the very tips of branches to pick fruit. They can do this by hanging one-handed from the merest twig whereas monkeys usually have to remain on the tops of branches. So, how can *beelow* males judge the capability of their neighbours to defend a territory? It's likely they do this by listening to their songs. A male can show his confidence and aptitude by singing for a long time and by performing vocal feats of trills and often repeated phrases. This is known as the 'handicap principle' with 'handicap' used in the horse-racing rather than the disablement sense. Thus, a really good horse can be loaded with weights, or handicapped, and still be expected to win against his rivals. Likewise, Sam takes so much time off from other important activities such as feeding and looking for new food sources, that no neighbour could regard him as a pushover. Sam would have to do something about his cough though. Males have only one song, and it also has a fatal attraction for females. Bachelor males sing for longer than mated males but adopt more normal habits once they have enticed a female to their home.

The relatively short female song has a simpler message: 'Any stray females thinking of wandering here hoping to get Sam needn't bother; I'm already here.' The female song is therefore also for defence, but defence of her chosen mate rather than of a territory. The dramatic visual display of leaping and leaf-tearing is a clear indication of the position of the caller and a reinforcement of her presence. As might be expected, spinster females away from their mother never sing.

Bess was leading the group up to the north of the home range, travelling by way of a small fruiting *sigeupgepi* tree. The fruit were like small plums with a deep-purple skin and reddish flesh, each of them an easy mouthful. They dropped a few which I tried and was pleasantly surprised at how palatable they were. They'd never sell well in a market, but given a few generations of careful breeding they could be vastly improved. The predecessor of the commercial plum varieties was the sloe, the tartest fruit you could ever taste. There wasn't enough fruit in the tree to keep the *beelow* there for long, and on they went diverging further from the path turning westwards. I cut across, joining Hanging Valley Trail where I had lost them some weeks before. In my notes for that day I'd written, '*Valley in front of me far too steep to get across.*' I had eventually managed to cut a way down, however, using small trees as hand-holds and as steps, and kicking my heels into the soft rock to make steps where trees weren't handy. The small stream at the bottom tumbled over a smooth, moss and lichen-bordered lip into a sixty-foot waterfall. No jumping down this one; there was nothing but loose scree at the bottom. The *beelow* had a route over this valley that was effectively level but I still had to climb up the other side and through a tangled treefall where I left bits of my shirt and shoulders on rattan thorns. I had phases of feeling uncontrollably guilty about cutting through rattan canes. They're commonly a hundred feet long and their existence is a never-ending struggle to reach the light at the top of the canopy. Cutting a small tree out of the way need not result in its death because suckers are often produced at its base or side shoots can grow. Cutting a rattan, or any other palm, spells inevitable death because the only growing point is right at the tip. I picked some thorns out of my shoulder, sighed, and pushed my way to the top.

I searched the canopy for the *beelow* and found Sam 'waiting for me'. As soon as our eyes met he set off down the other side of the home range, following Bess and Katy, giving me hardly a chance to catch my breath. I could hear his females ahead of us and it wasn't until we reached the flat area that we all stopped for a rest.

Bokkoi head

An hour later the only movement from them was an occasional scratch, performed not from the wrist or elbow as we would scratch but from the shoulder. It was a very gross movement but obviously satisfying. Suddenly the relative peace of the forest was shattered by a tremendous '*bagok-bagok-bagok-bagok gulugulugulugulugulugulu eeee-or*'. This territorial call of the male and female *joja* monkey was so loud that I could have believed they were only a couple of feet away but I couldn't see any sign of them.

I reckoned that the *beelow* wouldn't move for a while and so went down the path a short way to see where the *joja* were. There was a crashing of leaves in front of me and I saw an adult male *joja* leaping between two trees. I nipped behind some palms and peered out. Behind the male were the adult female nursing a grey-coloured, two-month old infant; a fully coloured juvenile; and a rather older adolescent. The female called, '*Go-go aak*,' quietly and went towards the male, with the infant carried below her belly rather than on her chest as I would have expected. The juvenile whistled and bounded after them, its long, glossy-black tail flowing behind it like a silk streamer.

Again the male bellowed out, '*Bagok-bagok-bagok-bagok gulugulugulugulugulugulu*,' and the female finished off with her donkey-like '*eee-or*'. As they called they bounded up and down on the branches making their tree sway. Bess left her tree and came towards me and the *joja* family, and the other *beelow* came too. I was poised over my notebook ready for some interesting interactions to begin but nothing happened. At one point the two families were completely mixed but the eight animals studiously ignored each other, passing like ships in the night.

I moved to one side to follow the *beelow* again and the male *joja* caught sight of me. He gave a single harsh '*bagok*' and leapt erratically above me. He branch-bounced, *bagok*-ed and tore in irregular circles around me. While he did so, the rest of his family skulked off and I continued to watch the male. This was a typical distraction display in which an adult tries to lead a potential predator (or research scientist) away from its young. We've all watched blackbirds hop around on a lawn feigning a broken wing hoping to lead one away from the whereabouts of its nest. Eventually the male *joja* rushed off, glancing back over his shoulder to check whether I was following.

The *joja* and *simakobu* (the other leaf-monkey on Siberut) are peculiar among monkeys in the Old World in that they live in monogamous family groups just like gibbons. Other monkeys live in the

Toga examining an orchid

Uma of the Sabulau clan

Beelow and Joja dance

A woman fishing for prawns

Aman Uisak divining the future from heart veins

Mother and baby *joja*

Simakobu

familiar troops, consisting of one or a few males and many females with their offspring. Indeed, there are eighty-three species of Old World monkeys but only two, the *joja* and *simakobu* leaf-monkeys of the Mentawai Islands, live monogamously. This begs the question, 'Why?', but seeing as how the inevitable answer is, 'It's anybody's guess', it's worth playing around with some ideas. It could be suggested that since many of the features of Siberut seem to be so relatively ancient, maybe the peculiar social system is primitive and that other species have evolved to become polygamous. This is unlikely because males of any animal species would prefer to have as many mates as possible so that they can produce large numbers of their own offspring. As long as the females are protected and able to find enough food for themselves and their offspring, they'll usually accept this state of affairs. Males therefore need to be driven to monogamy and there has to be a good reason for them to accept it. These good reasons fall into two categories. The first is that he has absolutely no choice. This would happen, for instance, if males and females were so thin on the ground that no more than one of each sex would ever meet at the same time. The second reason is that the male has no real choice. This would occur if the carrying capacity of the habitat was low so that the male couldn't defend an area large enough for more than two adults. Alternatively if the female produced twins or very large infants that the male had to help care for, a second mate would reduce his own offspring's chances of survival. With gibbons monogamy is probably of the second type – a result of the small, scattered food sources they

84

depend on. To ensure an adequate supply of these, it's best to live within a fixed area, that can be learned well, with as few others of the same species as possible.

Could any of these arguments apply to the *joja* and *simakobu*? At first glance it seems unlikely because the forest on Siberut is superficially very similar to that on the mainland where their close relatives live in troops. There are the same species of trees growing to the same height in the same sorts of places. It has been suggested that Siberut leaf-monkeys live in small groups as a response to hunting pressure but their relatives on the mainland and elsewhere have been human prey for far longer than has been the case on Siberut and they also have to deal with nocturnal cats. Also, the *joja* at least shows features that are typical of monogamous animals such as duetting and the adults' being very similar in size and colour, both of which would have taken longer to evolve on Siberut than the period of Man's presence on the island.

The only realistic conclusion is that there must be something weird in the forests on Siberut that makes these monkeys monogamous, and the most likely factor is that their food is less available than it appears to be. Notable features of the Siberut environment are the very high rainfall and the highly unresistant soils and rock. Together these result in minerals being washed into the rivers and so the plants would be expected to hang on to their hard-won nutrients with any means at their disposal. One way they can do this is to make their edible parts toxic or at least unpalatable and two groups of chemicals that plants are known to use in this way are tannins, that cause your teeth and gums to feel dry when you drink red wine or bite into unripe fruit; and alkaloids, that give the bitter taste to coffee. It is possible, then, that although the leaves certainly look abundant and eminently suitable as monkey food, many of them may be packed full of unpleasant tasting substances and thus the *joja* and *simakobu* have to live at quite low densities and therefore in territorial and monogamous groups. Maybe; until leaf samples are compared with those from the same species on the mainland no one can be sure.

By now, Bess and Katy were feeding on crimson nutmeg fruits, the seeds of which had splattered on my head at dawn. Sam was in the lower canopy heading for a *pola* palm tree that had ripe fruit hanging from it in huge bunches. He had to enter the *pola* by walking along its long, stout leaves, arms held high above his head for balance. The fruit were about the size of a tennis ball and he was very particular to pick only the reddest. They had thick stems and he had to twist them, yank them and try to pull at them with his teeth while his hands and feet held on to firm parts of the two-foot-long bunch of fruit. He then used his powerful long canine teeth to pull parts of the fibrous rind away before reaching the edible parts. I tried some time later to open some of these fruit with my *parang* and it was extremely hard going; I dread to think what it would have done to my teeth. After about six tearing bites Sam was able to scoop out and swallow the large seed and the thin bland pulp surrounding it. Was it really worth it, Sam? In fact, eating *pola* fruit can be a risky business because the fruit contain a salt of oxalic acid which, when eaten, forms minute, needle-sharp crystals by absorbing calcium from the body. Not only are the crystals very painful but the loss of calcium can cause muscular spasm and, in sufficient quantities, death. Such chemicals are to prevent animals eating the fruit before it is ripe, when the seed would not be viable, and so the levels of this protective chemical decrease as the fruit ripens. I never saw Bess or Katy eat *pola* fruit, and I wonder whether this was because their teeth weren't strong enough or because they had tried once, bitten into a slightly unripe fruit and vowed never to try again.

Sam on *pola* leaf stem

With no warning, Bess and Katy left their feeding tree and rushed past Sam. He swung out of the *pola* tree and was close behind them. They brachiated rapidly southwards over the smaller of the swamps in their home range, bounced through the tree Sam had slept in last night and carried on to the lower part of the West Ridge. Then, as suddenly as their three-hundred-yard dash had begun, the trees stopped shaking and there was no more noise of crashing leaves. I watched and waited, and it was several minutes before I found them apparently snoozing at the top of a bare tree; Sam on one side, Bess and Katy on the other. It wasn't yet three o'clock and it seemed so unlikely that they had entered their sleeping positions for the night that I hung around. They had travelled about a mile so far which was about average for one day but there were still over three hours until dusk.

With predictable unpredictability, the sky was darkening with rain clouds and before long I could hear the noise of a million raindrops hitting a million leaves further north up the Paitan valley. The distant sound of thunder boomed across the valleys and I donned my hooded rain cape. Within moments, the rain had arrived and I heard Katy squeal with displeasure. It was a humdinger of a storm with thunder that would make the *joja* call sound like a whisper. Even through the plastic of the cape I could see flashes of lightning and in due course we would doubtless find trees that had been struck. By four o'clock the thunder had moved away but the downpour continued. The gibbons curled up as tight as possible, looking miserable with heads on knees, arms around bodies, not just to keep dry but to keep warm too. Unlike many mammals, gibbons don't have an inner coat of fine hair to act as insulation and so they present as little surface as possible to the elements, only occasionally shaking the rain drops from their fur and fluffing it up like a bird on a cold winter's day.

Trusting that the *beelow* wouldn't move again that day, I set off home along paths that had become the beds of little rivers. When I turned a corner and saw the Sibosua stream it was not the usual trickling brook but a torrent carrying surprisingly large bits of the forest along with it. As the path descended so it too became flooded and I frequently had to wade waist-deep, hanging on to vines to keep my balance. At the last river crossing before camp it was clearly a case for swimming so I held my bag of binoculars and notebook above my head and pushed off.

'Wet, eh?' was the only thing I could think of saying to Jane as I negotiated the slippery steps to the veranda. We were alone temporarily because Alan had gone up to the northern river basins with Aman Taklabangan for a couple of weeks to complete a survey of an area that was suffering the greatest onslaught from logging companies.

'Mmm,' she mumbled distractedly while carefully drawing graphs from her squirrel data. 'How do you fancy a swim in the Paitan while you're still soaking?'

'You're joking,' I laughed, 'I've only just swum across the Simamunengmonga.' I was persuaded, however, and we walked to the river which was now nearly overflowing its twelve-foot banks, stripped and dived in. It was somewhat like an exercise bath in that no matter how hard we swam, we still couldn't make any headway – any skill lay in avoiding the bits of tree rushing down towards the coast.

By the time Jane went over to Alan's house to cook supper, the whole clearing was awash, and when it was time to eat I unfastened our dugout from its mooring by the river and pulled it to the front door. Supper had been our time for listening to the BBC World Service, but just before Alan had gone to the north, our radio had ceased working. In most situations one would have thought simply that a transistor had worked loose, but in our remote position, our initial reaction had been, 'Is the World still there?' After that, the loss of our one-way communication with outside didn't worry us in the least – there was far too much happening with pigs being stolen, adulterers being accused and rattan prices going up. By the time we had finished supper, pots and pans had to be brought in from the kitchen platform, which was slightly lower than the floor of the hut, because they were in distinct danger of floating downstream. Even the clay fireplace was being washed away as the Paitan swept past. At eight o'clock the rain had stopped and we found that nearly four inches had fallen in as many hours. We used the dugout to get across to the other house, guiding our way using a pole and with Stumpy looking really worried at one end as he tried desperately but unsuccessfully not to get his feet wet. At nine o'clock the flood waters stopped rising just below the floorboards and we wondered whether this was good luck or whether Aman Bulit and his friends who had built the houses had actually known exactly how high floods would rise. In none of the four major floods we experienced in camp did the water ever quite come over the floorboards, except in the kitchen.

Jane and I were just drifting off to sleep when we heard a sound that we agreed was probably Stumpy chasing a rat across the bouncy palm floor. This explanation satisfied us until we heard the same sounds again; were we being overrun by rats? Jane peered out from under the mosquito net and couldn't believe her eyes. Stumpy was standing, tense and upright with his fur on end in the middle of the floor, and a large brilliant green pit viper was poised in front of him. I lifted the net up too, just in time to see the snake lunge forward like lightning, but not too fast for Stumpy who neatly sidestepped and in the same motion bashed the head of the pit viper onto the hard floor with a paw. That was the noise we had heard. The snake lunged again, but again it received a blow from Stumpy. The snake slithered towards our bed but Stumpy wasn't going to have that. He dug his claws into the scales of the snake's tail and pulled it back into the middle of the floor. Before it was able to turn round, Stumpy was out of the way and taunting it yet again. One false move on Stumpy's part and he'd be bitten with inevitably fatal consequences. We wanted to help him but with everything including our clothes out of reach we felt somewhat vulnerable. Then Jane had the idea of hopping onto the food box from the bed, and from there reaching across the window to the stout length of *ariribuk* wood used to fasten the shutters. She handed this to me saying, 'Go on, it's your turn now.'

'Thanks a bunch.' I tiptoed behind the snake and in my hesitation I almost lost my chance as the viper wriggled towards the rows of tall biscuit tins we used for storage. It was almost out of sight but Stumpy caught hold of its tail and dragged it out. I caught it just behind the head and pinned it against the floor. Stumpy was so excited he hit its head like a boxer with a punch-ball, and I had as much difficulty avoiding his claws as restraining the snake. When it was dead I preserved the head in formalin and gave the remainder to Stumpy; he nibbled at bits of it only to vomit it up again a minute later. We got back to bed again and heard Stumpy miaowing to join us. 'All right, just this once then, seeing as how you've rid us of lurking serpents,' Jane told him.

'I thought the stilts were supposed to make the house snakeproof,' I murmured sleepily.

'They probably do,' she replied with a big yawn, 'but I suppose the snake must have been washed against the house in the flood.'

'Maybe.' In fact, that wasn't the last we saw of snakes in the houses. Several times, and not always after floods, we disturbed snakes among the biscuit tins and Stumpy found a huge female pit viper peacefully shedding its skin underneath Alan's camp bed. Alan must have slept happily, only inches above her, for at least two nights.

Next morning I walked gingerly to the veranda, shook my tall boots and put them on before I did anything else. I shone the torch on the ground and there was no sign of the flood save a thin layer of mud over the grass. The rivers I had to cross to reach the *beelow* were higher than normal but the palest image of what they had been the previous afternoon. I settled at the place from which I had seen the *beelow* brave the rain, and waited. About ten minutes before dawn, Sam stirred, held his arms out like a furry black teapot and shook himself. He looked around at the sleeping shapes of Bess and Katy and then gave a high piping note with his mouth held wide open. There were a few other males singing but none had progressed very far in their songs. By dawn Sam had probably made his presence felt in the neighbourhood but he had hardly performed any vocal feats. When I later analysed the information on male singing before dawn, I found that not only were they very unlikely to sing if rain had fallen in the night, but they hardly ever sang if the temperature was below seventy-one degrees. The dividing line was remarkably precise. In order to sing, *beelow* have to raise their heads and by so doing they expose their very sparsely haired throat and chest. Below seventy-one degrees, the costs of losing heat from their bodies presumably exceed the benefits of singing. In fact, every morning on which Sam failed to sing before leaving his night tree could be explained by either cold or rain.

My *beelow* had been active for barely nine hours the previous day and had foregone the final daylight hours when they could have eaten. Most animals use the whole breadth of the hours available to them even if they do rest in the middle of the day, but gibbons do not fit this pattern. The most plausible reason is linked with their comparatively rare, scattered sources of fruit and the complexity of the fruiting cycles. By the law of averages there will be periods when there is little or no fruit available within a gibbon family's home range and if it were normally necessary for them to fill all the daylight hours by feeding, they would suffer considerable hardship when fruit was hard to find. They therefore have a hedge against the bad times. Similarly, a family could doubtless exist in an area of less than eighty acres but it maintains and patrols the boundaries of a larger area to retain the exclusive use of it, just in case suitable fruit should become rare and the larger area is essential in order to find sufficient food.

They headed off down the West Ridge and fed in a fig vine right on the edge of the swamp. I stood at the end of my bouncy bridge across the peaty Situi river and watched them. A few days ago Bess's belly had appeared rather larger than I had remembered it and this morning it was looking more swollen still. Her nipples also seemed to be slightly elongated and her chest hair might have been even sparser than usual. It was an exciting thought that she might be pregnant: this seemed quite likely since Katy was about two years old and with a seven-month gestation period, sexual behaviour would probably have been occurring during the period that these *beelow* ran away at the first sight of me. It would be extremely interesting to have a new member of the group and would provide the opportunity of studying the developmental stages of a young *beelow*.

They ambled across the swamp, but didn't stop. Indeed, they usually travelled across the swamp without pausing because the density of suitable fruit trees was very low and few of the trees were tall and ideal for singing from or resting in. As they climbed up a vine near Summit Two the sun came out, and it was predictable that they would sunbathe for a while to warm up after yesterday's chilling storm. Several feeds and rests later Bess and Katy were travelling to the Glade by way of Summit Three and Sam was taking an alternative route to the west. By splitting up occasionally like this, each individual stands a greater chance of finding grubs to eat that haven't been disturbed by preceding gibbons and they are also more likely to find a source of fruit that is beginning to ripen. Suddenly, Katy started screaming at the top of her voice and dashing around in the tall dipterocarp trees. She seemed terrified and, as she rushed off towards the headwaters of the Tolailai river, I followed her. I couldn't make out what was wrong – could she have been stung by a bee or maybe Bess had bitten her in a fit of temper? Maybe she had seen a snake or perhaps a scorpion?

For at least seven minutes she screamed horribly and Bess didn't follow her as she usually did. I guessed that Bess had taken a shorter route to the Glade and when Katy eventually arrived there I assumed that the whole family was together again. It was always difficult to tell the adults apart when they were in the north of the study area because the trees were so tall. It was also the middle of the day and the sun was just about directly overhead, making the animated black blobs against the glaring sky look the same. By the middle of the afternoon I still hadn't positively found Bess again and so when I knew Sam had found somewhere to sleep below the Fig Trail, I walked back to Summit Three again. The tree in which I had last seen Bess was tall and broad and she could have lain on any one of the major boughs and remained out of sight. I looked on the ground below, thinking she might have fallen, but couldn't find anything out of the ordinary.

Next day, I watched Sam and Katy leave their respective sleeping trees and sure enough there was no sign of Bess. It seemed likely that she was becoming more sedentary and less sociable prior to giving birth. I knew that a wild white-handed gibbon female had been reported to opt out of her group for a few days when she had given birth. This was presumably to avoid leaping through the trees while the infant had too little strength in its arms to hold on to her. Almost every morning for the next ten days I spent the morning or some of the afternoon with Sam and Katy but caught no glimpse of Bess. Katy was having to grow up very quickly; she was forced to sleep on her own for the first time in her life and she received no comfort from Sam. He was belligerent towards her, particularly when it came to feeding in the same tree, and she spent a good deal of time whimpering pathetically to herself.

At the end of the second day without Bess, I arrived home to find Aman Bulit cutting down some saplings behind the house.

'*Anaileuita*,' I called, 'what're you doing with those?'

'These? I am taking their bark,' he replied. 'Wait a short while and I shall join you.' A few minutes later he stopped and we had a cup of coffee and some *patara* fruit together. These forest fruit grew on a small spiny tree behind our house and Jane had made pies, tarts and an excellent jam from them.

'What's the bark for?' I asked.

'For this,' and he demonstrated how by flexing the saplings, each produced two seven-foot lengths of broad, hard-wearing straps. 'They are to be attached to our large baskets,' he continued, 'for we are making sago in the swamp over there. Perhaps you would like to come over and watch.'

'Fine,' I agreed and picked up the camera before following him across the Paitan, which was not quite deep enough to wet our shorts or loincloth respectively, and into the sago swamp a short way downstream. Although Aman Bulit must have weighed at least as much as me, despite being considerably shorter, he sank a fraction of the distance I did into the peaty mud. Also, the sparseness of hairs on his legs meant that when he pulled his legs out they looked relatively clean. As for mine, the particles of peat wrapped around the hairs, giving the impression of a rather unpleasant skin disease. Unfortunately, sago palms grow only in swamps and if just a small area is suitable when the first 'sucker' is planted, the coarse, matted roots soon form a dense matrix which extends the area of waterlogged soil, thereby allowing more suckers to thrive.

Soon we reached the start of a pathway of felled sago trunks, some more rotten than others. Walking along these was really easy because they were about a foot and a half across and very rough. There were sago palms and little else as far as the eye could see, which wasn't actually very far, all covered in creepers and moss. The leaves were enormous, about six yards long, and they grew out of the top of the trunk like arms craving to touch the clouds.

'We've got to walk in the mud again here,' Aman Bulit smiled, 'but you'll find sections of sago bark just below the surface.' Sure enough, by putting my feet into the watery holes of his prints, I made contact with invisible solid matter. Shortly we arrived at the small clearing Aman Bulit and Aman Taklabangan had made where their selected sago palm would fall. I estimated its trunk to be about fifty feet tall, not exceptional on Siberut but roughly twice the height they grow to in Sumatra and Malaya. The difference may be because most of the sago swamps on Siberut are surrounded by forested hills that provide a steady flow of nutrients to the lower lying areas, whereas on the mainland sago palms are rarely in such positions and often have to compete for nutrients with biannually harvested rice. This particular one was just about to flower and was full of starch. Sago palms flower only once, using up all their accumulated starch; having fruited, they die.

'*Anaileuita aleh*,' I called to Aman Taklabangan, who returned my greeting. 'You don't mind if I watch you working for a while do you?'

'*Teelay*, no; but you can help me lay leaves along here first.' He pointed with his chin along the direction the tree would fall. We positioned long, heavy sago leaves at right angles to this, because otherwise the palm would probably disappear below the swamp surface.

Aman Taklabangan cleared the trunk of moss and orchids, and then began to cut away on the side facing the clearing. Whereas normal trees have their hardest wood in the centre, palms have an extremely hard bark surrounding a relatively soft pith. He used a cutlass-shaped *parang* to cut through the bark but then switched to a heavy axe to remove the fibrous white pith. When it was cut half-way through, Aman Taklabangan and Aman Bulit stood back discussing modifications and refinements to the cut and once these had been executed, we all stood behind Aman Taklabangan while he made cuts on the back of the trunk just above the big wedge. The sago palm mumbled, then groaned, and then roared as it tumbled forward exactly where they had planned, its leaves swishing through the air.

'*Teelay, ai'ei*, there it goes,' we all yelled in excitement as the ground shook below our feet. We heard several other '*teelays*' from up and down the Paitan and the whole valley would soon know that Aman Bulit and Aman Taklabangan were making sago.

We sat on the trunk to rest and Aman Bulit said, 'Making sago never used to be as hard as this, you know.'

'How come?' I asked.

'Well,' he began in his story-telling voice, 'once, in a time before sago grew on Siberut, there was a man and his wife who had one child, a girl. One day the mother said to her daughter, "*Elay, maynita paligagara*, let us go fishing." So they went fishing, the mother with a net and the daughter with a net. They fished and they fished and they fished until they had caught a great many fish and prawns. Meanwhile the daughter was getting hungrier and hungrier and hungrier and just as they were about to leave for their *sapo* she cried out, "Eeeeeee, I am hungry, eeeeeee." She sobbed, "I want, I want to eat sago, eeeee."

' "What do you mean you want to eat sago?" said the mother mystified. "What is sago? I have never heard of sago. Be quiet!"

' "Eeeeee, but I want to eat sago, eeeeeee," the daughter cried again with tears running down her cheeks.

' "I have told you we have no sago. I do not even know what it is. Be quiet!" But still the daughter cried. The mother took her hand and led her along the long path to the *sapo* but when only half-way there the daughter began to sob again. "Eeeeee, I want to eat sago, eeeeeee."

'The mother fetched some *bairabbi* fruits from the nearby tree. "Here," she said, "eat these *bairabbi* fruits if you are hungry."

' "No," shouted the daughter loudly, "I want to eat sago, eeeeeee. I do not want to eat *bairabbi*."

' "You cannot eat sago; there is no sago; there never has been sago!" the mother shouted back angrily. When the two of them reached the *sapo* the daughter was still crying out for sago and the father heard the noise and came to see what was wrong. "*Teelay*, Daughter, what is it you want?" he asked.

' "Eeeeeeee, I want to eat sago. I am hungry, I want to eat sago, eeeeeeee."

94

' "What do you mean you want to eat sago? What is sago? I have never heard of sago. Be quiet! Here, have a hen's egg," he offered.

' "Eeeeeeeeee, I want sago. I do not want a hen's egg, eeeeeeee."

'For the rest of that day and the following day the daughter cried continuously for sago. This so infuriated the parents that, at length, the father took hold of the daughter and threw her out of the *sapo* onto the ground where she died. That night a sago palm began to grow out of the body of the daughter and by the next noon it was full-grown. The parents were amazed to see it but even more surprised when it spoke to them. "This is the voice of the sago palm. Yesterday I was your daughter; now I have changed into a sago palm. You could not give me the sago I wanted to eat because there was no sago. Now listen carefully and I will tell you how to collect sago flour from my trunk." The sago palm did just that; it described how they should lean a bamboo container against the trunk and cut a hole in the section of the trunk nearest the ground. Sago would then simply flow out into the bamboo. When more flour was required, the next section should be pierced and so on up the tree, building a ladder to reach the top.

'The parents did as they were told and they had abundant sago to eat. Then, one day, the husband went hunting and the wife went to collect sago flour on her own. She carried a length of bamboo into what had now become a sago swamp and when she laid her bamboo against the sago trunk and cut a small hole above it, sago flour immediately flowed into the container until it was full. She started to walk back to her house but tripped up on an exposed sago root. She fell flat on her face and the container was smashed, spilling all the flour onto the mud. She hit the sago palm angrily, swearing at it, and went off to fetch another container. When she returned, she was still angry and cut another hole. Nothing happened.'

'That's right,' interrupted Aman Taklabangan, 'no flour at all came out of the trunk.'

'So,' Aman Bulit said quickly, regaining his usurped position as chief storyteller, 'she got angrier and angrier, hitting the tree, and eventually chopped down the sago palm. The other sago palms were watching her and were horrified at what she had done. They vowed there and then never to give up their flour so easily again and not to let women extract it. Instead, men would have to do the work although the women could help them.'

'*Teelay*, some help! It's us who have to sweat,' grumbled Aman Taklabangan.

The next day, work on the sago palm began in earnest and I joined them in the afternoon after checking that Bess hadn't reappeared. When I arrived at the clearing, Aman Bulit and Aman Taklabangan were sitting on either side of a section of the trunk where the bark had been peeled away. They held opposite ends of a curved narrow plank, reminiscent of a medieval instrument of torture, with dozens of nails driven through it. They pushed and pulled this between them, and grated pith accumulated at their feet. The wives gathered this into a pile and chopped it with *parangs*. When sufficiently fine it was piled into large baskets and carried to the *pusaguat* or sago-making place. Before the men left the clearing, however, they hammered a heavy circle of wood into the end of the trunk so that wandering pigs wouldn't be able to feast themselves on the pith. Pigs would sell their souls for a bite of sago and Aman Taklabangan made quite sure that the wood was fixed firmly. It is interesting to consider that if two of those large wooden discs had been placed side by side with a stout sapling joining them through the holes, there would have been the beginnings of a cart. The fact that the inhabitants of Siberut have never exploited the principle of the wheel is, however, absolutely no discredit to them. It simply reflects the fact that any Siberut cart would have been bogged down in mud two minutes after the start of its maiden run.

The *pusaguat* is part of the swamp where the slow flow of water has been dammed to form a pool of deep, brown water. Over the pool had been built a platform which supported a huge sieve made of wood and coarse fibres from the leaf bases of *pola* palm. Below the platform were screens of sewn sago leaves which diverted water falling through the sieve into part of an old dugout. This sloped into a complete dugout anchored firmly to stakes driven into the peaty mud.

Jane arrived at the *pusaguat* just as Aman Bulit was emptying the contents of a basket into the sieve. '*Anaileuita eyrakku*,' he called to her, 'catch this,' and he tossed her the empty basket. He climbed into the sieve and pulled up a bucket of water. The bucket was made of sago leaf bases, had a conical base and was attached to a long bamboo pole. Anyone who has tried filling a conventional bucket from a well will appreciate the neatness of this bucket's construction. He started to dance on top of the flour, like someone treading grapes, while pouring water over his feet. He trampled, moved, rubbed and crushed the sago pith until the water flowing into the dugout was rich with starch and looked rather like dirty milk.

Bai Bulit arrived from the clearing with another basket full of pith and joked with her husband about how slowly he was working. 'Why not let Tony try to do better,' she suggested. Everyone looked at me expectantly and so I washed my feet in the swamp water and climbed into the sieve with Aman Bulit. Personally, I was quite pleased with my efforts, concentrating on squeezing the soggy pith between my toes and remembering to tread near the edges as well as in the middle. Aman and Bai Taklabangan appeared from the clearing and laughed uproariously at the unexpected sight and my admittedly limited success. The others joined in and I gave up after inadvertently breaking the bamboo pole in two by pulling the bucket up at the wrong angle.

The sago palm was eventually cut into eleven sections and it took six days (two hundred man-hours) to process all the pith. By this time the dugout was full to the brim with sodden flour and the next stage began. The women collected a large number of sago leaflets and Aman Taklabangan sorted out the longer ones. He sat in a little shelter and prepared to make *tappris*. These are double-walled containers used for storing sago flour, and are made by carefully interweaving and sewing the leaflets. When ready for use they are cylinders about a yard high, ten inches across and closed at one end.

Aman Bulit scooped off the surface water in the dugout and then stepped gingerly into the sticky flour. Bai Bulit brought over the first *tappri* and Aman Bulit dropped clods of flour into the open end. When the *tappri* was nearly full it was shifted to one side ready for Aman Taklabangan to *pasipokpok* it. This involved beating its sides with a heavy piece of wood to settle the flour and to remove excess water. Fourteen *tappris* later, all the flour had been packaged up, making a total volume of about half a cubic yard, equivalent to one quarter of the pith volume.

Although sago flour shouldn't be wet when stored, it must be kept moist to retain the flavour. The traditional practice is to bury the *tappris* in the mud below the swamp until they are required. This may seem a ridiculous notion but decay processes there are slow if not entirely absent because of a lack of oxygen, and sago flour will keep for at least a year without signs of deterioration.

Sago flour contains little except starch and water with protein accounting for barely one per cent of its weight – much less than for rice and many root crops. This lack of protein has led the authorities to suggest rice as an alternative staple for the people of Siberut. This is, however, falsely argued for a number of reasons. Firstly, the climate on Siberut is unsuitable for rice as there is no truly dry season to ripen the seed heads. Secondly, rice suffers from a multitude of insect, bird and mammal pests, against which people who cannot afford pesticides or other means of defence are helpless. Thirdly, sago takes comparatively little time to process, leaving ample time for family members to fish in the rivers and sea, to tend their pigs and chickens and for the men to hunt occasionally. At the present population levels there is no shortage of 'wild' protein on Siberut provided people have the time to gather it. Fourthly, sago can be converted into animal protein by feeding it to pigs and chickens and through a process known as *mutamara*.

A few weeks later Aman Bulit came to see us, saying he was going to *mutamara*. Puzzled, Jane and I followed him to the sago swamp and there we met up with Bai Bulit making wooden wedges from a small sapling. The plan was to climb a sago palm and split its trunk from the base to the top while it stood. Aman Bulit tied some flexible rattan cane loosely around his waist and the tree, hung an axe, a large *parang* and some thin rattan from his loincloth and started to climb. He untied the axe, held it above his head and swung it into the bark making a vertical split two feet long. He called to Bai Bulit who threw him a wedge and he forced this into the crack. Using his long *parang* he cleared all the moss and vines from the trunk above him, climbed a little higher and with perfect accuracy split the trunk some more. Bai Bulit tied wedges to the thin rattan cane which he hauled up. Higher and higher he climbed, joking about what he might cut off if he fell.

The purpose of all this was to attract large black and red weevils by the smell of the exposed pith. Sago has no pests to speak off, but if the weevils are allowed through the bark they will lay eggs and the grubs from these then start to eat their way through the slowly fermenting pith. After another seven weeks we met up again with Aman and Bai Bulit and his Aunt Lasui. By now the split in the sago palm had become slimy and brown and by putting your ear to it, it was possible to hear the strange sounds of hundreds of little mouths munching away at the pith. Aman Bulit felled the tree and as it crashed down it split open revealing some of the long, bloated grubs. Jane and I were shown how to sort through the soft, pungent pith, and armed with buckets the five of us set to work, collecting all the wriggling, white, squeaking grubs we could find. Just one hour later, we had gathered over 1300 of them weighing twenty-six pounds from the thirty-seven-foot trunk. In the meantime two of Aman Taklabangan's children had arrived, and delighted in horrifying us by eating the grubs alive. Adults usually eat the grubs roasted; they are rich in fat and protein, taste not unlike pork and must be one of the most easily collected protein sources around. At first the thought of eating them was quite repulsive, but in response to a dare from Aman Bulit, I managed to swallow one quickly without chewing. By the time we left Siberut, we were buying them from people, having them as grub fried rice, grub soup, grub and corned-beef fritters, and grub with chilli dip. Goodness knows what would have happened if we'd stayed much longer.

On the thirteenth morning after Bess's disappearance, I was with Sam as he picked some *popokpok* fruit. These look like large grapes and grow on short stalks straight out of the trunk and branches. As he climbed to the right of the tree he noticed a line of ants walking along one of the boughs. He climbed onto the bough close above them, leant over and wiped the furry back of his hand along the moving column. He brought his hand to his mouth and licked off the ants and went fishing again. Soon, of course, the line of ants was disrupted and they swarmed all over the bough, but Sam found the densest patches and used longer sweeps of his arm to pick up the teeming insects.

Suddenly we both heard the introductory notes of a female song coming from the west. I'm sure that our thought processes were the same; first we convinced ourselves that the sound was coming from a neighbouring home range, but then it dawned that it was far closer than that, probably just a few hundred yards away on the West Ridge. Sam left his confused ants and rushed at right angles to the Main Trail towards the swamp. I followed as best I could and scarcely two minutes later we both arrived at a treefall on the West Ridge. I crouched behind a bushy tree and saw Sam watching a female sitting in an open-crowned *alibagbag* tree with spoon-shaped leaves. The introductory notes were, to put it bluntly, pathetic. She wavered about all over the place and suffered from the same dry throat as Sam sometimes did but in her case after singing for only two minutes. When she started her first great-call, the 'finest music made by any land mammal' became a shambles and I swallowed a laugh. It was as though she knew the tune vaguely but had left the music behind some tree and couldn't quite get it right. Even Katy could have put her to shame. She gave up, thankfully, after three tries but the message had been released – 'Sam has got another female'.

The driving force behind all strategies of breeding is the desire by each adult to leave as many of its own offspring as possible to the next generation. So, if this female was to become Sam's new mate, they would first have to test and ensure each other's future fidelity. From the point of view of females, if they all withheld sex for a number of months until the male proved able to maintain and protect his territory, then it wouldn't be worth his while to mate repeatedly, desert, found a new territory, attract another female and go through the vetting process again. A female and her offspring need to live within a secure territory if the rearing is to be successful. It is sensible for the male too to withhold sex because he has to check her past affairs and ensure that his potential mate isn't already pregnant. What would be the benefit to him of helping someone else's infant to grow in his territory? Some recent gibbon studies have found that the first mating of a new pair is, indeed, delayed. Once the first conception is achieved they are both paired for life for, by the time that desertion by either parent is not detrimental to the first offspring, another will have been born and the parents are caught in a fidelity trap.

We can only guess at what had happened to Bess on that day Katy had been shaken so badly. She could have died of some obstetric complaint, or perhaps a python lurking in a tree had caught her as she passed. But that would have surely caused her to make at least some noise and would Katy have screamed so much? We shall never know.

I wondered whether Sam would like this female. I moved to get a better view of her but she saw me, hooed excitedly and took off. In her hurry she brushed past Sam who was sitting facing me and then looked behind her with an expression of total confusion. Didn't Sam realize that there was a human down there and didn't he know it wasn't safe to just sit around? She gave up and raced off again. Sam picked a vine shoot and chewed it. He raised his eyebrows as he peered at the disappearing shape of the female and the way he looked back at me was as though he were saying, 'Women, they're crazy.'

Chapter Six
Simalegi Survey

'The area around the Simalegi river is probably where man first lived on Siberut,' Aman Bulit told me as he brought glasses of hot, sweet, black tea onto his veranda. 'Up in the north, people still believe that the first person to set foot on the island was a pregnant woman who had been adrift on a raft. She gave birth to a son shortly afterwards and he grew to be tall and strong. When he became a young man, his mother gave him her ring and told him to go off and find a bride whose finger it fitted. He travelled the island for many years but found no women at all. Eventually he came across his mother in the same place where they had parted but neither of them recognized the other. The ring fitted her finger perfectly and so they were married, and everyone today is supposed to be descended from them.' Aman Bulit chuckled, obviously not giving much credence to the story, and swallowed some of the scalding tea.

'To start with,' he continued, 'people stayed in just the Simatalu and Simalegi basins in the north-west, living together in peace, but then, maybe five hundred or maybe a thousand years ago, clans began to make furious raids on each other and people travelled southwards through Siberut to the southern islands.'

Some of this folk history is borne out by anthropological findings. Judging from their language and physical characteristics, the Siberut people almost certainly settled first in the north-west of the island. This happened roughly three thousand years ago. They are defined racially as 'proto-Malay' having a Neolithic culture with some Bronze Age influences not influenced by Buddhism, Hinduism or Islam, and this race of *Homo sapiens sapiens* was probably the first to arrive in Indonesia. It is impossible to tell exactly when the migration through the islands began, but man did not arrive on South Pagai until about three hundred years ago. Evidence of the migrations is also seen in the presence of a small number of clans, such as the Sagaragara along the Paitan river, who own almost no land. They were always late in claiming virgin land and now are somewhat despised, living on other clans' land and blamed for thefts and other misfortunes.

The early settling in the Simatalu and Simalegi basins is thus of great anthropological (and possibly archaeological) interest. This and their relative inaccessibility make us feel that the area might be suitable for inclusion in the new nature reserve. I had walked to Sirisurak in the hope of persuading Aman Bulit to accompany us as a guide, but both he and Aman Taklabangan were in the middle of felling a new clearing and couldn't spare three weeks off work.

I sat sipping the now-bearable tea talking to Aman Bulit about nothing in particular, when a young man approached whom I recognized only vaguely. He wore shorts and a clean but holey vest and a broad but nervous smile that revealed chipped incisors. His hair was swept straight back accentuating his high forehead. 'Er, M-Mr Tony,' he stammered softly, 'I-I heard from some other people that you, you, well, that you were, er, looking for someone to take you to, er, Simatalu and Simalegi.'

'That's right. Can you help us?'

'Well, yes. That is, er, if you like.'

He was Tupilaggai but on joining the Catholic church he was renamed Primus. He later changed his allegiances and became a Protestant called Ohn, but many people knew him as Aman Doiroishi. In the end we settled on Ohn as it was the easiest to remember. He had once walked to Lita, the Simatalu village closest to our river basin and was confident he could find someone there to help us go further. We fixed the delicate matter of wages, although he didn't seem too worried about receiving money, and fringe benefits such as the goods we could buy for him in Padang next time we went. We arranged that he should come up to camp four days later and that we'd leave for the west the morning after that. He oozed an impish sense of humour and a quiet charm that suggested travelling with him would be fun.

Before going on survey, Jane had to close down thirty squirrel traps that she had been setting for the previous eight weeks. She was hoping to catch *jirit*, *loga* and *soksak* and keep them captive temporarily so that she could, among other things, discover the contexts in which different calls were given and mark individuals so that they could be recognized again. Every third day she had rebaited the traps with pieces of ripening banana, fresh coconut, charred coconut or a concoction of ground peanuts and sago but almost every day they were empty. I say 'almost' because she successfully outwitted two butterflies, one frog, three lizards, four red forest rats – and Sidney.

Sidney was a fine, black-bellied *loga* who was caught not once, not twice but three times and always in the same trap. She kept him for several quite useful days, marking him first with a bright orange collar, and when that slipped off two weeks later when he was caught again, Jane trimmed the fur from a section of his tail. All this was an attempt to make him recognizable at a distance in the forest, but despite long and careful watching we only ever saw him in his trap. In fact, even with another hundred traps due to arrive in the next few months, it was becoming fast apparent that after the survey she'd do better to reposition them and use them for the red forest rats and thus begin a complementary project on other small mammals.

Jane and I were already awake when the six o'clock alarm rang. Pigs vomiting below our bed and stormy rain had given us a bad night and it was still drizzling when we had a wash in the river. Ohn and Alan were in the other house getting dressed and finishing the packaging. Last night had been the first time that either Jane or Alan had met Ohn and although conversation slackened at times it was an enjoyable evening, full of bonhomie, plans and excitement. We felt nothing but admiration for him; he had agreed to leave his family, pigs, sago and other comforts for three weeks of travelling in regions unknown to him with three foreign strangers whom he'd doubtless heard could be impetuous and difficult at times.

'Right then,' I said to Ohn after breakfast, 'I think we ought to try to reach Simabuggei by nightfall.'

'Er, yes; er, do you know the way?'

'Don't you?' Alan asked.

'Er, no.'

We all burst out laughing; we were laden with rucksacks, standing in a fine drizzle in the middle of our clearing, intent on travelling but our newly-employed guide didn't even know the way for the first day. However, we knew we had to go upstream to start with and with Ohn leading we kept as close as possible to the Paitan. Within an hour we had reached the Sagaragara clan's *uma* near which we luckily found some of the young men felling trees. They gave Ohn rapid directions on how to negotiate the major ridge between us and the Bulu river and, when he was confident, we began the climb. Teitei Sibajet, the ridge to the west of the Paitan was about five hundred feet high; not exceptional, but sufficiently steep and slippery to make the thought of the three other hills that had to be scaled before nightfall rather daunting. At the other side of Teitei Sibajet Ohn guessed which path we should

103

follow but at one point he became more nervous than usual and made us take a long detour around the path. We kept asking him why, but all he'd do was mumble something about being shy. It wasn't until months later that we discovered that his father-in-law lived very near the path we had been following and that husbands on Siberut don't engage in unnecessary contact with their fathers-in-law because of a form of incest taboo.

Just before dusk we arrived at Mr Tseng's shop which looked amazingly orderly since a couple of months' worth of copra had just been taken to the mainland. Alan had had a pretty uncomfortable day filling his hand with *ariribuk* spines, his shoulders with rattans, twisting his temperamental knee and being troubled by acute athlete's foot. We debated whether he should go home but he persuaded us that he'd be all right.

The westward path from Simabuggei began as a scramble over smoothed boulders in the bed of the Uselat river, but after an hour it rose and we proceeded to climb over the island's major divide. The path was excellent, the best forest path we ever found on Siberut, primarily because it was used so regularly by hosts of Simatalu people hoping to sell goods to Mr Tseng. What bridges there were, were very sturdy and well maintained although Alan very nearly fell off one trunk onto the rocks below. At the highest point of the walk, a grassy glade, we sat down and chewed on some old sago and fried dried fish that Jane had located in one of our rucksacks. After eating, Alan lay back looking drawn and tired from the effort of having to use his bad knee and soon went to sleep.

Not long after we had left the shop we met a couple of young Simatalu women carrying copra for sale to Mr Tseng. They had now managed to catch up with us and they pointed at Alan stretched out on the ground, remarking to each other in utter amazement about his length and asked if he were dead. They giggled a while and then joined us for some sago and a smoke. When everyone felt refreshed we woke Alan and the women argued briefly over which one of them should carry his high-frame rucksack down the hill to the Popangi river. Once there, they made him lie in the bottom of their dugout and poled him gracefully downstream while the rest of us walked and slipped our way to another dugout moored to the bank.

Not only do languages and tattoos vary between river basins, but dugout design changes too. Here, instead of having relatively steep sides, they had scarcely more curvature than a dinner plate. We thought we had learnt how to sit in a Siberut dugout (it honestly does need learning), but we shipped water repeatedly and even Ohn was finding it tricky to manoeuvre; judging from the expression on his face, however, he was having a whale of a time. The river was very different from those we had seen on the east of the island; instead of being languid and meandering it was swift and shallow with many shoals covered in white water. Every so often we had to drag the dugouts over the rocks and these were probably the major reason the dugouts weren't deeper.

Two monkey tattoo designs

The women came from Lita, the village Ohn had visited some years before, and we stopped there to spend the night. Ohn went off to find someone who could take us through Simatalu and beyond, and after a few hours emerged from the houses with a man named Aman Jenga Kerei. He looked almost Italian in his features with loose curly hair and wide-set eyes. He had an irrepressible and infectious smile which revealed that a couple of his chipped teeth were missing and it was to prove invaluable over the following weeks. He spoke almost no Indonesian but understood our ungrammatical Saibi version of Mentawaian and spoke clearly and slowly in his own language for us. We tried to discuss wages with him but he kept laughing and saying he didn't really want anything at all and was quite happy just to travel with us; we postponed the subject. His full name was rather a mouthful so we adopted the more informal Simatalu address of '*Bolaikta*', which literally means 'our friend'.

While we were seated on the floor of the village chief's house with dozens of villagers smoking us out of tobacco, I brought out my animal cards with which I was trying to document the different regional animal names. We had tested out our drawings on Aman Gogolay and some of them had to go through several drafts before he would at least hold the card the right way up. It was an entertaining

diversion on survey because as much time was spent telling each other
tales of the last time they had seen each of them as deciding which was
the usual name in the region. The primates, deer, rats and civet cats
went well but when I brought out the drawing of a false vampire bat,
the house went silent apart from a few whispered *'teelays'*. I pressed
some of the men for its name but they said they'd never seen it. 'Surely
you . . .' I started, but received a kick in the back from Jane before I
could finish.

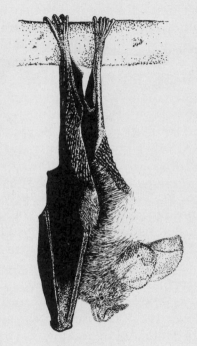

False vampire bat

'*Bajo*,' she whispered urgently, 'Ohn says the bat has a strong *bajo*
in this area.'

I quickly covered the picture, not to mention my embarrassment,
and proceeded to the next card. Even so, the air seemed to have
become colder and the spirit of joviality had gone. As soon as I had
finished, people excused themselves and left. Next morning the inci-
dent seemed to have been forgotten and we made an early start. We
weren't able to go where we had originally planned because of a lack
of paths but even facts like that provided us with some information on
land use. Bolaikta's wife, one of the women who had helped the day

before, paddled us all upstream a short way, gave us some sticks of sago and set us on the path towards the Torajan river. Even this path was relatively little used and Bolaikta had to put his back into clearing a way through the knotted undergrowth. This was no hardship because he had recently purchased a new *parang*, of which he was inordinately proud, and which he seemed to have spent most of the early hours of the morning sharpening outside the house where we had slept.

At the top of the first hill was a most unlikely sight – a large lake. 'This is the only hilltop lake on Siberut,' Bolaikta proclaimed grandly with a preposterous grin. Aman Bulit had told us about this lake months ago when we first mentioned the possibility of walking in Simatalu. 'Take care,' he had said, 'take care, because the lake has no bottom. There used to be hundreds upon hundreds of *toulu* terrapins in it. But then a huge beast, something like a crocodile and a python, arrived and ate all the terrapins. Be careful if you should ever see the lake because the beast will be hungry and may try to eat you.'

Aman Bulit had obviously never seen the lake. If it had no bottom then the sago palms some enterprising soul had planted in it had mighty long trunks. Also, judging from the din, either the beast had starved to death or it didn't like the taste of frogs.

By mid-afternoon we were walking along a major ridge above the Torajan river hoping to reach an *uma* Bolaikta knew by nightfall. Suddenly the sky darkened and it started raining, really raining, and we hurried our pace along some knife-back ridges so narrow that our feet were on either side of the divide. By the time we reached the flat ground by the Torajan, the river was a swirling torrent and the storm showed no signs of abating. We agreed to make a shelter and within minutes our two guides had constructed a little hut from stout bamboos, covered with overlapping banana leaves each of which was three yards long and a couple of feet wide. Banana leaves were also laid on the ground to provide a floor and there was ample room for all of us to change out of our wet clothes. Only a few more minutes passed before our guides had managed to conjure up a roaring fire of split bamboo at one end of the hut and it was soon surrounded by steaming loincloths, shirts and shorts. Ohn was still splitting old bamboo for the fire when suddenly he exclaimed, '*Teelay, sikaobbuk, teelay!*' He cupped his hand over one of the bamboo halves and beckoned us over. As he slowly raised his hand we saw a minute bat nestling just below one of the bamboo nodes. I took it out gently and we looked in amazement at

107

its tiny piggy eyes and vicious little teeth. Its head and body were less than two inches long and it must have weighed less than a quarter of an ounce; a wonderful example of biological construction at its finest. Although we could hear only the faintest squeaks he was evidently bellowing at us as loud as he could, so I took him out into the warm drizzle and tossed him into the air like a piece of irate gossamer.

We had planned to reach some form of habitation each night, but on the offchance that this proved impossible, we carried several days' supply of rice and dried foods sent to us by British companies. There was a selection of soya mince, mashed potato and sliced beans, all of which could be cooked in lengths of bamboo. Our guides tucked eagerly into the steaming food piled on a banana leaf from the roof, but remarked that it all tasted rather strange. After supper Ohn attempted to learn the few words Bolaikta already knew in the Simalegi language. Poor Ohn clearly tried very hard but he didn't fare too well. He just couldn't stop himself laughing when he heard the peculiar Simalegi accent Bolaikta imitated and the peculiar, totally different vocabulary we were going to use over the next week.

Next morning the river flowed unhurriedly, defying any thought that it could have been a deep and dangerous torrent only hours before. We tried to walk along the sides of the river bed but this was easier said than done. The spate had brought down quantities of extremely fine mud, like sediment from a china clay or cement works upstream, which clung tenaciously to our legs. It was rather like walking in thick custard or blancmange as the mud wobbled for at least two feet around each footfall. When we stood too long in one place, we were sucked down several feet and had to summon the help of others to pull us out.

By the afternoon, we were all extremely tired and our guides were running out of the dried palm leaves or *boboirot* they used to roll their cigarettes. During one of our stops we heard voices and Bolaikta rushed off with renewed energy to find the people. A little while later he returned along the river looking as pleased as punch, poling a dugout, holding the bamboo pole in one hand and a bunch of *boboirot* in the other. His welcome news was that there was a groups of *sapos* not far away where we could stay. Ohn cut himself a bamboo pole too and we made rapid progress downsteam. Ohn was in a splendidly giggly mood, so pleased was he to get out of the mud, and even falling backwards out of the dugout as it struck some submerged wood didn't dampen his spirits.

After settling in at an empty *sapo* where Bolaikta assured us we'd be welcome, Alan and I walked with Ohn to the other *sapos* and *umas* to count monkeys skulls and to meet the people. One toothless, lined, skinny old man we introduced ourselves to looked momentarily surprised at our arrival and then burst out laughing as if to say, '*Teelay*, not "whities" again!' We felt we must be tagging along behind some tourist party until it transpired that the last Westerners he'd met had been Dutchmen making a map of the island. That was in 1931.

We decided to stay two days in the area because of the wealth of skulls to count and because it was impossible just to arrive, count skulls and leave. Instead, we offered tobacco, chatted as best we could, and were presented with mounds of boiled bananas and coconuts; only then was it reasonable to ask permission to count the skulls. None of us really expected such hospitality and skull counting was abandoned for the day when bananas started coming out of our ears. We virtually had to hide in the forest to avoid being offered any more bananas, because the surfeit of them was beginning to play havoc with our insides.

In the twelve houses we visited, we found 430 monkey skulls. Unfortunately sheer quantity can't be used as an indication of hunting pressure because skulls from an old house are often, but not always, moved when a new one is built, and so there is no reliable way of estimating either the time over which the skulls have been collected or the number of people they have fed. The skulls are useful, however, in estimating the relative hunting success on the three monkeys. In most of the river basins we visited, we found that about half the skulls were *simakobu*, a third *joja*, and a fifth *bokkoi*. In Simatalu the ratios were significantly different and consistent between houses; just over half the skulls were *bokkoi*, a third *simakobu* and a fifth *joja*.

What could have caused the discrepancy? In some basins, *bokkoi* are caught in rattan and bamboo cage traps which can be very effective, catching up to twelve of them at once. But these traps are not used in Simatalu. We saw more *bokkoi* crashing through the trees away from us in Simatalu than anywhere else on Siberut and so it could be that they are simply more abundant there. If so, it is important to find what factor or factors might cause the difference. Simatalu is unique among the populated river basins on Siberut in one important respect; rattan is not cut for sale because boats cannot enter the harbour mouth to take it away. I saw *bokkoi* eat rattan fruits several times in the study area and other macaques have been reported to eat rattan fruits elsewhere in South-east Asia. Whether this is the whole story or not is impossible to say, but it is possible that commercial rattan collection makes less food available for *bokkoi* and so affects their abundance.

During the day while Alan and I were counting skulls Jane had met up with the couple who owned the *sapo* we were staying in and their three young daughters. The chubby, friendly one-eyed wife was incredulous and pitying when Jane was pressed to explain that, no, we hadn't left our children behind anywhere, we had none. They all went fishing and before they left they changed out of their cloth skirts into skirts of shredded banana leaves which cause little disturbance while walking in the water and so don't worry the fish. Jane was surprised when the three daughters, the oldest of whom was barely seven, headed up a different tributary with their mini-nets and mini-banana skirts, to return at the end of the day with a genuine contribution to the household; bamboos holding easily as many prawns, crayfish and fish as their mother had caught.

Just as we were all thinking of turning in for the night, the lamp began to flicker as a swarm of flying termites flapped clumsily through the *sapo*. There was great excitement; the lamp was carried to the veranda in front of the *sapo* and a large platter full of water placed next to it. The children were woken to share in the feast and as the termites fell into the water or burned their wings in the flame, so they were snatched up. It was supposed to be children's food but Bolaikta, never one to miss the chance of eating, was pouncing on the crawling bodies with the best of them. Wriggling termites was the one Siberut food I could never bring myself to eat.

It was a privilege to live and work among people who were largely unmaterialistic but this trait had its drawbacks. We couldn't find anyone who was willing to guide us to Simalegi through the forest; despite generous offers of remuneration. The more usual route was by river and a number of people were willing to help paddle us by dugout, but this would have told us nothing of the nature of the forested hills. There were hushed stories that anyone venturing where we wanted to go was bound to die.

So, the five of us set off hopefully in a north-westerly direction armed with a compass, copies of the old Dutch map, a large bundle of sago cooked in bamboo and a couple of chickens packaged together in a tight, plaited palm-leaf shoulder bag. As far as Ohn and Bolaikta were concerned we could have ventured into the forest with no food at all as long as we had some slabs of tobacco. One of the agreed perks of being a guide was an unlimited supply of tobacco and Bolaikta seemed to smoke the sweet-smelling leaves constantly. When they were in company, however, they acted as though the shreds in their tins were all they had for the next month. Ohn had given us strict instructions about the amounts we should give to people on different occasions. We were apt to be over-generous and as soon as we were alone in the forest again, his nervousness evaporated and he scolded us for our mistakes. Tobacco was far more use than money when travelling and the slabs we carried were our wealth and insurance. Both Ohn and Bolaikta were always very careful not to let anyone outside our party see how much tobacco we had or where it was kept, because it would have caused any number of unwanted problems.

A combination of precipitous waterfalls, rain, and ridges not running the way we wanted slowed us down, and, by the time it was necessary to build a shelter, we were not convinced we'd left Simatalu. Our suspicions were confirmed next morning when we walked down-stream and met a family who, from their tattoos and language, were clearly from the Simatalu basin. We were, however, very near the major divide into Simalegi and after some discussion the man of the house where we had paused to rest offered to take us into the basin and on to the nearest village. First, however, he insisted that we have a meal with him and his family in their spacious *uma*. The menu looked conventional enough until we noticed a mound of thick, gelatinous, grey-looking spaghetti that the mother told us was *towek*. It looked like long boiled worms and was evidently a real treat and a great local delicacy. They weren't in fact worms but the world's largest marine, wood-boring, bivalve molluscs. They are collected by the laborious procedure of felling a *tumu* tree, cutting it into convenient sections and then towing a raft of sections to the Simatalu estuary. The raft is kept there for a few weeks for the *towek* larvae to start boring their way in and then towed back to the houses upstream. I would have expected the developing molluscs to have been killed by this change to fresh water but they keep on growing and when they are required for food, a section of trunk is broken open and the *towek*, some of which measure a metre in length, are extracted. Inhabitants of other river basins feel that the Simatalu people have rather strange habits, not only because they eat *towek*, but also because they are the only group to eat pythons and giant rats.

Towek were scarcely our all-time favourite food but we managed

111

to eat a few of the rather tough siphon tubes and filled up with sago. We had hardly finished our last mouthful when everyone jumped up and started to leave. This was all very well but after ten minutes I had vicious stomach cramps and had to stop the party for a while to let the *towek* and everything else settle.

It was a longer walk than we had bargained for, and we didn't reach the village of Simalegi Kailaba until after dusk. The people there were wonderfully friendly, carrying our packs for the last few hundred yards, children squealing with excitement in front of us. We were taken to the village chief who was very welcoming and adamant that as soon as we had washed and changed, we should eat at his house. As it turned out, the whole village seemed to eat at his house but he didn't baulk at feeding them all with at least some rice and a delicious-looking sauce of coconut milk, turmeric and chilli. The yellow of the turmeric disguised the quantity of chilli that had been ground into the sauce but it was so hot that none of us, not even the guides, could eat it. Behind us the chief's wife was forcing the violent sauce into the mouth of her screaming baby. The poor child was obviously in great distress but in theory the sooner it could take strong chilli, the sooner it would maintain a relatively worm-free intestine.

Conversation buzzed around us but I could understand only about one word in twenty. I leaned over to the guides to ask them what was going on but not even they had a clue. They looked almost as much like fish out of water as we did, and everyone in our party sat with half-closed eyes longing to get some sleep before the next day's exhausting trek.

Our initial plan had been to travel in the east of the Simalegi river basin, crossing between river valleys and eventually arriving at the coast. However, the village chief could find no one who was free to take us and so we resigned ourselves to paddling down to the coast for a couple of days making occasional sorties to either side. We had imagined the Simalegi basin to be relatively backward owing to the irregularity with which boats can enter the harbour, and its distance from Siberut's two administrative centres. We were surprised, then, when it turned out to be one of the most 'civilized' basins we visited, with most of the people living in large villages by the main river surrounded by extensive coconut groves and other fields.

Before leaving Simalegi Kailaba we were offered a meal and grate-fully accepted. We sat cross-legged around the dishes, said grace, and started to unwrap the sago parcels. The stew in front of us was dark brown and as people started to take bits of meat out of it, so one particular piece worked its way towards me and emerged out of the gravy. It was a little furry hand with the fingers seemingly reaching for the edge of the plate in a hopeless bid to escape. I had learned enough of the language to ask what it was. '*Satettet*,' or '*simakobu*' came the reply. Jane, Alan and I looked at each other not knowing quite what to do but when someone mercifully took the hand we tried, and even enjoyed, the meal. The meat was gamey, extremely tender and lean, and the rich gravy was delicious.

The meandering journey to the coast was slow and uncomfortable because the dugout Ohn had hired was really too small for all of us and our packs. No matter where we stopped, Bolaikta was always first out and by the time we found him again, he'd be chewing someone's stick of stale sago, bouncing children on his knees, and would have found a long-lost relative who'd be arranging for some fresh coconuts to be cut open for us to drink. The majority of these were delicious but some were slightly fizzy and provided the most gorgeous, thirst-quenching drink imaginable. Our guides offered around their *boboirot* leaves, but before we squeezed into the canoe again, Bolaikta had spirited up a handful of spares.

The mouth of the Simalegi river was worth waiting for. Two thickly-forested cliffs descended virtually to the water's edge and there was none of the usual muddy sand and merchants' houses. We drank in the sight, smell and sound of the small breakers ahead of us for a short while before walking up the coast to the new village of Simalegi Muara. One thought that struck us, however, was that if any of us became ill here, it was a good six day's journey back to camp let alone to medical aid. Simalegi Muara had been in existence for only two months but already had a few permanent houses and wide paths marked out. The village chief, a middle-aged man called Markus, had his house closest to the river mouth and he greeted us warmly for he said he'd heard we'd be arriving soon. How anyone could possibly have got to him before we did was a mystery.

We told Markus how we had planned to walk through eastern Simalegi, emerging at the Takmaeruk river to the north of his village and that we'd like to walk at least some way along it.

'It is fortunate you did not walk along the Takmaeruk by yourselves,' Markus said seriously. 'The name of the river means that it is no good. No one lives along it; no one ever has, although some people have built shelters for when they cut rattan in the hills nearby.' This helped to explain some of the difficulty we'd had finding guides.

'Why don't people live there?' Jane asked.

'*Sikaoinan*, crocodiles live there,' he replied, 'and they have killed some of the village children.' Later in the afternoon, anxious to see at least the mouth of this mysterious river, we strolled north along the sandy coast stopping half-way in a coconut grove owned by Markus. Ohn shinned up one of the trees and we stayed clear while he twisted ripe nuts off their stems and sent them plummeting to the ground. Then we all sat in the shade and drank pints of the sweet milk. Apparently, the people now living at the coast had once lived some way up the Simalegi but the ground became 'tired' and their crops wouldn't grow, so the whole village moved. We had heard similar stories in other areas and the consequences were almost inevitable if people settled in villages and cleared large areas for planting. I also talked to Markus about the crocodiles. It had been ten years since a crocodile had hurt anyone, and the major reason for this was that ten years ago a leather dealer had come over from the mainland and shot twenty-five adult crocodiles in a day along the Takmaeruk. Crocodiles didn't exist officially on Siberut; that is, they had never been collected or sighted by anyone who wrote reports, and we needed to obtain first-hand evidence of their existence. Crocodiles couldn't be given any form of protection if they didn't exist on paper.

On the way back, Ohn found a round *Callophylum* fruit and started kicking it along the beach. Alan tackled him and an energetic game of football began in which Ohn tended to win most of his tackles as he was still wielding his eighteen-inch *parang*. It was wonderful to be able to run on the hard white sand after ten days of walking along muddy paths in the dense forest and sitting in the confines of a dugout.

Markus put the large tin plate to his lips and slurped the final drops of fish juice remaining from the selection of fresh fish we'd eaten for supper. He wiped a soft, bouncy length of baked sago around the plate and after swallowing this looked up, satisfied. '*Kawat*,' he said, 'now we shall sit outside and talk.'

As we got up from the mats on the floor, about ten sleek, well-fed cats moved in, searching for any scraps we'd dropped during the meal. Markus's wife cleared away and by the time she was ready to join us on the veranda, there was hardly any room left. The whole village – man, woman, boy and girl (and a few dogs and cats) – had come to sit there

A *Silokgob'buk* lizard

A small field house

A new frog species

Toga at six months

Inside a field house

too. Everyone was dressed in blue or red tartan sarongs which they gathered around them as protection against the cool sea breeze and the ever-present mosquitoes.

The sea, fifty yards away down the beach, shimmered beneath the rising, young crescent moon and the phosphorescent glow from the foaming breakers appeared and disappeared as they sprawled over the sand. An orange warmth on the veranda spread from a couple of small flames serving the dual roles of cigarette lighters and lights but this wasn't sufficiently blinding to prevent us contemplating the sparkling stars. Markus saw me looking out at the moon and said, 'There used to be a time when the moon was full and round every night.'

'What happened to it then?' Alan asked from behind us and people shuffled closer to hear what Markus was going to tell us.

'Well, there was a time when the moon and sun always followed each other across the sky. Both of them had children but when the sun and her children shone they scorched the earth and burned the people when they went to the rivers to bathe. The moon felt sorry for the people and planned to trick the sun.

'The moon smeared a red juice around its mouth and the next morning she told the sun how that night she had eaten all her star children and how extremely good they had tasted. She managed to convince the sun to do the same and the sun immediately started to eat all her children too. She agreed that they had, indeed, tasted very good.

'As the sun fell and entered the realm below the earth, all the stars came out and she began to realize that she had been tricked for there above her were all the moon's children. This made the sun very angry and she took her *parang* and cut the moon until it became smaller and smaller. This made the moon very ill and she went to her turmeric fields. The maggots from her wounds crawled to the water and these are the *simaru* fish that swim up the rivers for a few days every month. But every time the moon recovers the sun remembers her trickery and hacks pieces off it, causing the moon to return to the turmeric fields.

'The moon was unable to retaliate in the same way because the sun was too hot and so impossible to approach too closely. But she managed to cut pieces out of the sun's outline and that is why the sun never has a clear outline and why we sometimes see rays sticking through the clouds.'

'Beautiful,' Jane exclaimed, and she thanked Markus expansively. In a world of 'inexplicable' phenomena and naturally inquisitive minds, such stories give reassuring meaning to the universe. Jane and I had twice seen and been puzzled by the amazing monthly shoals of millions upon millions of *simaru* fish swimming up the Saibi. That they were the ailing moon's maggots was a far more charming explanation than any we had been formulating.

The following day we gave our boots a well-earned rest and our toes some freedom by walking barefoot down the coast. The broad white beach stretched far away into the hazy distance and when the sand became too hot to bear we let the warm, deep-blue sea lap around our feet. The beach was dotted with shells of all shapes and sizes and colours, from lobsters, cone shells, varieties of cowrie, clams, scallops and, most impressive of all, the nauptilus. Bolaikta collected a number of these and added them to his already straining rucksack. He was an inveterate collector and even half a green football, presumably jettisoned from a ship on the high seas, was included in his booty. He put it on his head but the stiff breeze lifted it off. So, he scoured the ground and eventually came up with a short length of plastic string half-buried beneath the sand. With this fixed to the edges of the football it stayed in place and he strode ahead with a preposterous royal dignity spoilt only by the occasional giggle.

At the top of the beach was a vertical, yard-high wall of sand marking the upper limit of the tides. Whereas most of Siberut's coastline has been planted with coconuts, there was nothing above the beach wall here but natural coastal vegetation. In patches there were almost pure stands of conifer-like she-oak with delicate, long drooping branches, but between these stands were thickets of leathery-leaved *Barringtonia*, tall and stilted pandans with harshly-toothed leafblades, and hibiscus bushes, some of which bore large yellow and purple flowers. In a few places, on the very edge of the vegetation facing the sea, were little groups of beautiful lilies bearing large flamboyant white flowers with spindly white petals looking like some weird spider *in extremis*, and at intervals along the beach monstrous *Callophylum* trees sprawled heavily from the beach wall onto the otherwise bare beach. The wind and the rough seas that extend almost uninterrupted between these Siberut beaches and those of East Africa, and the absence of a protective coral reef don't allow mangroves to grow on the west coast in anything like the profusion they grow on the east. We found little pockets of them sheltered behind small sand bars at the mouths of rivers but their general scarcity probably accounted for the fact that the majority of loincloths in this region were a cream colour rather than the red produced by soaking the barkcloth in mangrove sap.

116

We were walking past a group of *Barringtonia* trees when, out of the corner of my eye, I saw some branches move violently about twenty-five yards away. Bolaikta had picked up the movement too and we both stopped dead in our tracks. The others halted sharply and we all watched in acute surprise as a large, dark-grey male *simakobu* jumped cautiously between two trees. We saw his mouth move as he turned to face us, but we could hear nothing above the constant roar of the waves behind us. He dropped out of sight but then others started to follow along approximately the same route he had initiated. Jane, Alan and I became tense with curiosity when we had counted four, including an infant, cross the gap (four being a common size for a *simakobu* group) but could see two more preparing to jump. Then another climbed into position; and yet another. Eight *simakobu* in all, with definitely more than two adults. All of them were the common dark-phase; roughly a quarter of all *simakobu* are a splendid creamy-gold colour.

Bolaikta forestalled the expected avalanche of questions by telling us immediately that large groups of *simakobu* are common in the northern river basins and this 'normal-monkey' social organization was later confirmed by a Japanese zoologist working near the north-east coast. But why should *simakobu* be polygamous in the northerly basins but not in the majority of Siberut? I have already explained that a male animal has to be forced to accept monogamy, so the observed polygamy here is due either to the relevant pressure being released or to an additional pressure causing polygamy to be the lesser of two evils. I suggested that toxins in tree leaves could have caused the peculiar mating systems of *simakobu* and *joja*. The obvious way out of the problem is to propose that such toxins aren't present in large quantities in the northern trees but the soil and rainfall do not seem to be noticeably different from the south. Of the available alternative explanations, the one I find most convincing is that the large groups are a response to prolonged hunting pressure. The regions where polygamous *simakobu* are found are, after all, where man has lived and hunted for the longest time. *Simakobu* we had observed in our study area were ridiculously maladapted to human predation. They are not terribly aware to start with, but when they actually do see a human they frequently just sit tight and hope that you'll go away. A hunter rarely does, however, and Ron Tilson recorded how *simakobu* could be caught by felling the tree in which they were hiding. This has resulted in the *simakobu* becoming the rarest but most hunted Siberut primate.

The large group size of *simakobu* in the north can be beneficial in two ways; first, there is a greater chance that at least one of the others in your group will see a predator before you, and second, there is less of a chance that the one to get shot will be you, considering that a hunter will rarely take more than one animal at a time. It would make a fascinating study to investigate the ecology of the large and the small groups of *simakobu* and to determine in what ways they differed.

Our destination for the night was the mouth of a little river called the Gobjib. Our old Dutch maps showed that this flowed from a series of small brackish lakes and we wanted to see whether these were still surrounded by forest or whether monotonous coconut groves had superseded it. As we approached the Gobjib we saw a thin plume of smoke rising through the trees, and then being buffeted inland by the strong wind. Bolaikta went on ahead to prepare whoever it was for the onslaught of visitors. The rest of us followed his footprints inland and we soon came across him seated in a shelter talking through a mouthful of cold sago to a couple of hunters and one of their wives. Bolaikta discovered that the hunters were brothers who lived in Simalegi Kailaba, but every few months they walked over the hills and came down to the Gobjib area to hunt. There was a rustling in the bushes behind us and the second wife appeared carrying the furry, headless carcass of a *simakobu* in one hand and its dripping, cleaned organs and intestines in the other. She dumped the latter unceremoniously in a blackened cooking pot which already had the singed, split head in it and then built up the fire, the smoke of which we had seen earlier. Her husband held the carcass over the flaming fire to singe off its hair, for nothing is ever skinned on Siberut: what's the point of wasting potential food? An hour later we were presented with a tea-time snack of lukewarm sago and a steaming gut-and-head soup drunk from a communal coconut spoon. We behaved like good guests and expressed appreciation.

Jane and I wandered down to the beach at dusk to watch the sunset. The sea now looked grey and dull but it acted as a neutral backcloth for waves catching the remaining coppery light in their foamy crests. The sky was streaked with reddening clouds, and to the north, deep yellow rays of light struck at gathering storm clouds apparently trying to keep them at bay. We could hear a high piping call and then we watched a large wading bird, the greater thick-knee, flap slowly past. It was such a glorious setting and such a wonderful evening that I shouted with the sheer joy of being alive. We started to

Simakobu

run hand in hand along the wet sand, just like the advertisements for shampoo or rum, until Jane's face contorted and she clutched at her stomach and ran off the beach into the bushes. The head-and-gut soup came to be blamed for a great many afflictions.

After a supper of boiled *simakobu* I asked the hunters, through Bolaikta, whether there were any crocodiles in Gobjib. Yes, they said, they were everywhere around the lakes. Would they be prepared to take me out to find some? Yes, if we took torches and if we could go armed. With bows and arrows? No, with crocodile spears. One of the hunters prised a viciously hooked, heavy metal spearhead from the thatch. Bolaikta pointed to an eyelet in its side and said, 'You tie the end of a rope to this. When you get close enough to the *sikaoinan*'s head, you push the spear through the skull. This makes it dive and you keep hold of it with the rope.' The frightening technique sounded rather like a method of whaling and the spearhead could well have been traded from a whaling vessel years ago.

As the evening drew in, the tales of heroism and bravery on crocodile hunts of the past turned to tales of death, maimings, and taboos. When I again broached the subject of finding crocodiles that night I was given a list of excuses as long as a sago trunk and we retired

119

to our mats and mosquito nets rather disappointed but at least unscathed.

At dawn the next morning we paddled ourselves to the lakes in two small dugouts, and as we glided into the first and largest, fell silent. The wind was still and the early sun shone golden on the dark verdant forests that almost enclosed the rich mahogany-coloured water. Suddenly the tranquillity was broken by the clamour of *joja* calls followed straight away by the barking of a male *bokkoi*. We manoeuvred the dugouts to the edge of the lake and a *simakobu* gave its donkey-like call somewhere within the forest. A female *beelow* on the adjacent hill began her great-call, filling the natural amphitheatre of the lake with her song and we looked at each other not able to find the right words to express the wonder we felt at being in this wonderful place. We disturbed some huge flying foxes roosting in the nipa palms that bordered the lake and they flapped lazily around us before disappearing from our sight. The nipa roots were covered in *bokboket* whelks and startlingly pretty fishes darted in and out, occasionally rising out of the water and causing the lake's only ripples. Nowhere on Siberut had we experienced such an abundance of life and never had we heard all four primate species call in such a short space of time. We left knowing that this area would have to be included when we eventually proposed the boundaries of the new reserve, and thinking with rather mixed feelings that this part of Siberut might one day become popular with tourists.

The animals on Siberut will never be a powerful tourist attraction. They generate a specialist interest founded on their unique origins and the dramatic ecosystem they occupy, but most of the species are difficult to find and see, and have relatives on the mainland that are generally more accessible. A water hole by the veranda of a Siberut hotel of the future would attract little other than muddy feral pigs, possibly a nervous nocturnal deer or two, some dragonflies and countless mosquitos; not royal tigers, proud lions, graceful antelope or mighty elephants. Instead, when tourists go to Siberut, they will be seeking the luxuriant forest, the sand, coral and sun, the traditional people and their culture. This could be catered for in two ways.

First, there is the adventure holiday for which people pay to experience the feeling of 'camaraderie in adversity' as they suffer on arduous walks from leeches, floods, scratches and blisters. They could meet the local people on their own terms and would intrude on any one family for only a couple of days at the most. This would be a low-impact tourist invasion requiring no hotels or other infrastructure; just good guides and planning. By varying the routes taken, it could financially benefit a wide spectrum of local people directly, although the whole operation could be run on a relatively low budget.

Secondly, there is the tourism known picturesquely as 'high quality, low quantity', which means there would be only a few, richer tourists present at any one time. Hotels, traditional craft centres, maybe even an airstrip, would be necessary and there is scarce likelihood of there being any real beneficiaries save a few entrepreneurs from the mainland. People from Siberut would probably end up as waiters who, twice nightly, would dance and sing traditional songs. This second type of tourism will inevitably be adopted in some form somewhere on Siberut and as much restraint as possible will have to be shown in the design of hotels and other necessary buildings, and in the handling of the profits.

The rest of the day was spent walking further south along the beach, crossing several tracks made by female turtles hauling their bulk up the beach to lay their eggs. At least, we think that's what made them because Bolaikta, still wearing his ridiculous football hat, made almost perfect replica tracks at their side by shuffling his splayed feet sideways down to the sea in a manner reminiscent of Charlie Chaplin.

From talking to people and examining shells in houses we found that three turtle species visit Siberut's coasts: the three-foot hawksbill, the four-foot green turtle, and the six-foot leatherback – the world's heaviest reptile. All these are in danger of extinction the world over from direct killing on nesting beaches or inshore, accidental netting at sea, and nest plundering. We hope that some day a turtle specialist will evaluate Siberut's shores as a potential turtle sanctuary but until then we were intent on proposing total protection for turtles on and around the island and including a proportion of the turtle beaches within the reserve boundaries. Turtle meat was clearly eaten by some coastal clans but nowhere could we find a group that would truly suffer if deprived of it.

Five days before we finished our three-week survey we were hailed from a *sapo* surrounded by nodding coconut palms. '*Aleh*! do you have any medicine for a sick stomach?' a man shouted from the veranda, and in the same breath continued, 'I have a baby *beelow* here; come and have a look.' We had no suitable medicine but went to see the *beelow* all the same. Sure enough, clinging to a chicken basket was a spidery, black furry mass. It squeaked as we approached and held onto its basket for dear life. The man's son put a ripe banana next to the *beelow* and it licked at the soft fruit in a frustrated, ineffectual fashion. Jane put her finger in the infant's tiny hand and lifted it to her. Its face formed a deeply furrowed frown and it yelled like a beast possessed until it made contact with her body. Bolaikta watched Jane and his face moulded into an absurdly wide grin that clearly showed all the gaps between his teeth. It was obvious from then how the next hour would proceed but it was only the irreverent Bolaikta who let on. The *beelow* bit rather ineffectually at Jane's chest, and no sooner had she scraped some banana onto a finger than this disappeared rapidly into the gaping, hungry mouth.

With Bolaikta as an interpreter, we were told that the *beelow* had been clinging to a female he had shot and so I explained to the house owner that hunting of any of the primates was, in fact, forbidden by Indonesian law, as was keeping young ones without a licence from the Conservation Department. The man said he'd never heard of any hunting laws, which was more than likely, but then he changed the subject to talk about what we were doing in Simatalu, where we lived in Saibi, whom we had met and what we had seen. His wife brought out a platter of thick, cream-coloured boiled bananas with big, black seeds to catch the unwary, and we ate these while being ever mindful of the possible effects on our stomachs.

At length, the man asked us if we wished to buy the *beelow*. I explained that to buy it would be illegal but that we would be prepared to look after it at camp where we could provide it with milk. Without a slightly more natural diet the *beelow* was sure to die. Bolaikta tried to suppress his laughter, failed and lay on the floor burying his face in his hands. The man was obviously disappointed until Jane suggested we take the *beelow* but give some tobacco as recompense for some bananas and for the chicken basket to put the *beelow* in at night. This met with approval all round and an hour later we were walking through the forest again with Jane carrying the basket and me the *beelow*. It was so small and frail that Jane feared she might trip and squash it. I was no less likely to fall but took the infant all the same. The skinny beast was barely seven inches tall when seated and fitted neatly into a hand. Every hour or so we stopped and fed it but for most of the time it held onto my sweaty, torn shirt and slept. That night we put it in the chicken basket, covered it with a shirt and after a few minutes of screaming tantrums, it went to sleep. The next morning was rather depressing because there had been nary a squeak from it all night; a least we had tried. Uncovering the basket, I saw, as expected, a shape lying still in the bottom. The eyes opened blearily, blinked and then the tremendous shrieks began until I lifted it out and scraped banana into its mouth. There was life in it yet. The *beelow* was so young that it was quite incapable of moving around on its own and if put on the ground could barely lift itself onto its elbows, looking like a furry crayfish. We were still four days from camp and hoped that it could survive that long without milk.

Chapter Seven
Toga

When we eventually arrived back in Sirisurak, we went straight to Aman Bulit's house and a crowd soon gathered. Ohn somehow managed to condense the news from three weeks of exciting travel into about half an hour and the assembled masses seemed to be suitably impressed. Ohn's status in the village had clearly risen and he was more confident in himself as a result.

The *beelow* attracted considerable attention. Aman Bulit chuckled to himself and said with mock severity to Jane and I, 'What *else* did you two do while you were away? You have been married for nearly two years and produced nothing, and then suddenly, you have a child. It is more like Tony than you Jane, with its dark hair, but maybe its eyes are like yours!' The crowd nudged each other and laughed, adding their own comments. Aman Bulit leaned towards me in a confiding fashion, and in a ridiculous stage whisper counselled me, 'Look, we all know it is your first one, so you must remember to make sure that Jane drinks enough, it is most important. If not she may find it hard to suckle.' Peals of laughter erupted and Jane didn't know which way to look. People murmured about '*togara*' or 'their child' and from then on the young *beelow* became known as Toga.

'I suppose you know that a woman once did give birth to a *beelow*?' Aman Bulit continued. He knew full well that we didn't and so settled down to tell us the story, while everyone else, who had doubtless heard it a dozen times before, found somewhere to sit and yelps and miaows were heard as places were usurped.

'It was like this,' he began. 'Once upon a time there lived a man and his wife who had no children. Each day the wife looked after the chickens they owned; early in the morning she would go to the place where they were kept to give them food and to collect the eggs they had laid. Near to that place was a large rambutan tree with branches bearing the most wonderful red, juicy fruit. Each morning as she fed the chickens, the wife looked up at the fruit and said to herself, "If I were a *beelow* I could climb to the top of the tree and gather as many of those fruit as I wanted."

'One day when she was feeding the chickens she saw a *beelow* in the great rambutan tree eating the fruits which she desired so much herself. "If only the *beelow* would throw down some of the fruit that he picks, then I too would be able to eat them." The *beelow* heard the woman and threw fruit down for her to eat.

'The next day, when the woman came to feed the chickens, the *beelow* was again in the rambutan tree, and she called to him, "Beelow, Beelow". He replied, "I am here, what do you want?" "Oh, Beelow in the rambutan tree, throw down some fruit for me to eat." Hearing this, the *beelow* picked some ripe fruit and did as he was asked. Thereafter, each day as she fed the chickens, the wife would call to the *beelow* and he would say, "I am here, what do you want?" She would ask for some fruit and he would throw some down to her. After some time, the wife and the *beelow* fell in love and the wife became pregnant. The husband, suspecting he might not be the father, asked the wife why she always took so long to feed the chickens. But she did not answer, and went as normal to the rambutan tree to eat fruit with the *beelow*. However, the husband was cunning and one day he followed his wife to the rambutan tree, but out of sight so that she would not know he was there. As usual the wife called to the *beelow*, and as usual the *beelow* replied, "I am here, what do you want?" and as usual she asked him to throw rambutan fruit down to her.

'The husband became very angry and when the wife eventually left to go home, he took his axe and cut down the rambutan tree. He caught the *beelow* and cut off its head and put it in the chicken basket where the eggs were laid. The next morning, just before the wife went to feed the chickens, the husband said to her, "Be sure to look well in the chicken basket today." And the wife said that she would.

'When she arrived she saw that the tree had been felled and she called out to the *beelow*. There was no reply. She called again but still the *beelow* did not reply. Then she remembered her husband's words and looked in the chicken basket. There was the head of the *beelow*. Of course, she was very, very sad because she was in love with the *beelow* and she was still weeping when she arrived home.'

Aman Bulit's throat was getting dry so he persuaded Bai Bulit to stop knotting her large fishing net and go inside and boil some water for coffee. He continued, 'Now, the time came for her to give birth and the child she produced was the same size as a human baby, but was covered in long black hair as his father, the *beelow*, had been. The child was known as Sibulubulu or "the hairy one". As Sibulubulu grew up, he spent all of his time on the floor by the fire where he was thrown scraps of food to eat. Sibulubulu always looked listless and sick. However, when he came of age, he married a girl from the neighbouring clan and she came to live in a new house that was built for them. Sibulubulu was too lazy to build his own. And this was an unusual marriage because the wife had to do all the work whilst her husband sat about looking listless and sick. He was too lazy to do anything.

'Now, Sibulubulu had a secret: he could take off his hairy coat and underneath was a strong and handsome man. But no one knew of this, not even his wife who had to spend all day fishing and collecting food. Whilst she was working, Sibulubulu would take off his hairy coat, store it in a bamboo container in the roof and go to wash in the river. There he made himself beautiful, spreading coconut oil on his hair and body and indeed he was a fine man. But whenever his wife came back, there was Sibulubulu on the floor by the fire looking listless and sick.

'One day, while Sibulubulu was making himself beautiful he decided to go and help his wife in the fields, and the wife, not knowing who he really was, said to him, "I would indeed welcome any help, for my husband just sits at home looking listless and sick, and will not work." So, Sibulubulu helped her to gather the roots. Before they had finished, however, he excused himself saying he had to go back home and wash.

'It happened that not far away from Sibulubulu's house lived another man and his wife who had just had a new baby. The wife had to stay in all day to look after the baby, and one day she watched Sibulubulu take off his hairy coat, store it in the bamboo container and then go to wash. The wife was very confused and resolved to tell Sibulubulu's wife about it. So, the next day, the young mother followed the wife to the fields and told her what she had seen. She added, "Wait until he goes to wash and then I shall show you where he keeps the hairy coat. Then you can burn it because you will find him a very handsome man." Thus, when Sibulubulu went to wash, the women crept into the house and burnt all of his hairy coat except for a little bit; just enough for his eyebrows. From that day, Sibulubulu remained as a handsome and strong man who helped his wife in her work. *Alepaat*, that is it.' Aman Bulit sat back as Bai Bulit brought out the coffee. The other villagers felt it was a good story-telling and wandered slowly away after finishing their coffee.

126

We hadn't realized quite how tired the three weeks of travelling had made us until we got back to camp. We sat around in the clearing for two days, reading and writing, with barely enough energy to cook. During the long survey we had discovered that phrases like 'I can't go on' and 'I've had enough' have no meaning when there is no viable alternative. Our limited remaining reserves of energy were all used up by ministering to the needs of Toga.

She had small white sores on her lips and tongue, caused by a vitamin deficiency, and we hoped that these would disappear once she was drinking milk again. In fact, that was easier said than done. Not having a baby's bottle in camp, we had to improvise with the rubber squeezer from an eyedropper. The first mouthful of warm reconstituted milk that we managed to get down Toga was met with a frightful look of horror. Little of it was actually swallowed and she was soon searching around in full voice for a nice ripe banana. We had somehow thought that she'd be craving for milk but the process of *un*weaning her back onto a milk-based diet proved to be quite difficult. When she saw the bottle of milk approaching she would shut her eyes tight in the vain hope that it would go away. We had to adopt a system of bribery, starting with two pieces of squashed banana and one mouthful of milk, passing through one of each to one piece of banana and two mouthfuls of milk. Toga showed early on, however, that she was quite capable of counting up to two and for many days she continued to take milk only under great sufferance. After about a week her sores began to disappear and she began to put on weight.

From her size, weight, erupting teeth and behavioural development we reckoned Toga was about five and a half months old. Gibbons weigh roughly four ounces at birth and are weaned and at least partially independent of their mother at one year but do not reach full maturity until seven. Alan was going home to England soon, but Jane and I planned a further eight months on Siberut and so it seemed just possible that we would be able to return Toga to the forest, possibly as an adopted sister for Katy. We would have to wait and see.

The Conservation Department had earlier sent us another young *beelow*, Albert, which returned to the forest quite easily when Alan was alone in camp. Albert was probably about a year old when we first saw him and was accompanied by another orphan, an extrovert *bokkoi* called Rover. The two of them played together endlessly and if Albert wanted to stay close to Rover, he had no choice but to go into the forest and climb after him. Eventually Albert joined the camp's

127

Bokkoi

closest *beelow* group that had previously comprised an adult pair, their subadult son, and juvenile daughter. It took a few worrying days for Albert to become fully integrated, but it appeared that, at least with a *beelow* of his age, short-term rehabilitation was feasible.

Toga soon began to exert her personality and decided whom she would, and would not put up with, often for no apparent reason. I had carried her almost non-stop for five days at the end of the survey, and as a result she became hopelessly fixated on me and would never be happy for long with anyone else if I was in view. She was very content to be with Jane and would behave quite normally, but only so long as I was out of sight. With Alan the story was different; Toga hooed with alarm and, once able to crawl, rushed away from him in blind panic. Why she objected to him so violently wasn't clear, but it may have

128

been something to do with his bushy orange beard. When local men and women came by to sell fruit or chat, Toga stumbled forward on all fours, mouth open wide and teeth bared, hissing aggressively. If she actually managed to get hold of a stray leg or toe she would land a bite sufficiently painful to cause the poor visitor to retreat from the veranda. The one exception she made was Aman Bulit, possibly because she sensed the deep, warm friendship we had with him. He was disarmed by her charm, and was happy to have her clambering all over him or nestling in his arms.

Stumpy was very tolerant of Toga, being prepared (under sufferance) to play tag, and even to let her win now and again. When he was in no mood for such games, Toga's main recourse was to surprise him by lunging forward and fixing her teeth around his scrotum. Thus provoked, he occasionally lashed out at Toga, who thought it was a great joke, and when he managed to get free, fled into the forest. One day he didn't return. We had several times come across him some distance from camp, and at first assumed that he must have gone exploring. After a week it sadly seemed more likely that he'd met his match with a snake. In the hope of keeping the forest rats out of the house we eventually got another cat, a slim, dainty kitten called Musi. She was no substitute and wouldn't have said boo to a goose, let alone a rat. She disliked getting her grey feet muddy in the clearing and looked to us to evict itinerant snakes. Toga reigned supreme.

During Toga's first two weeks with us, pangs of hunger struck her every two hours or so through the day, and she communicated these by screaming shrilly. Between feeds she often took naps on a lap, or clutched a sweaty shirt if her minder was working. If I was in camp sitting by the table she entertained herself by jogging the end of the ballpoint I was using and trying with all her might to lift pencils and rubbers before dropping them through the floor. In the forest she was happy to be hung from one hand on a sturdy branch where she twisted and swung for minutes on end as long as Jane or I were close by. She didn't yet have sufficient muscular power to climb and if she brachiated at all it was always downwards. Occasionally she spotted a tasty-looking insect on the ground but she was as yet too slow and gauche to catch anything, and not strong enough to pull off the leeches that caught her.

Back in the forest after the survey, I sat in the dripping wetness waiting for Sam to wake and for the sun to rise. I slowly became aware of some scufflings in the dense stand of trees in front of me and when

Toga brachiating

the animals causing the movements started to produce chucking noises, I realized they must be *loga* squirrels. They graduated from simple chucks to a wide range of noises such as 'sweeet, siat, chuck, chuckle and bop' sounding more like a breakfast cereal than squirrels.

About half an hour before dawn, Sam began to sing; nothing clever, just enough to let neighbouring males know he hadn't died in the night. Dawn itself was met by the raucous duets of *joja* monkey pairs calling back and forth to each other from their night trees, and then Sam left his and fed alone in a small fig tree. Indeed, it wasn't until seven o'clock that Katy and Sam's new female, Manai, arrived on the scene. In the week after Manai had first sung her great-calls in Sam's territory, he had at worst ignored her totally and at best hooed at her as though she belonged to another species. His opinion of her didn't seem to have changed. Katy on the other hand had latched onto her as a substitute mother; but some substitute. I watched as Manai started to feed in a red-barked *popokpok* tree that Sam had just left, and when Katy moved closer to her Manai opened her mouth aggressively and flicked her arm in Katy's direction. Katy yelped in anguish and swung out of the large tree to sit dejectedly nearby, only to be replaced by a loudly flapping hornbill. Manai took singular exception to this unrequested intrusion and chased it out of the tree as it trumpeted in confusion. Manai was in a particularly cantankerous mood.

When I moved to get a different view, Manai saw me and was torn between two reactions. As a compromise, she stuffed a few more *popokpok* fruit in her mouth and then lurched hurriedly out of the tree, somehow managing to keep her balance on the loop of a slack vine, while looking over her shoulder and hooing at me. Her nervous-

Sam and Manai

ness made even Sam jittery and I nearly lost them as they tore off. Eventually Sam quietened down but Katy went off with Manai, presumably seeking solace and understanding. There was, after all, no real reason why she should stay around with Sam who largely ignored her, or with me who just gazed at her for hours and wrote copious notes.

Sam found a gaunt, dead bough on a tall *katuka* tree and searched among the crevices of its flaking bark for something good to eat. He grabbed a handful of moss, lichen and pieces of wood, and proceeded to sort through it by rubbing his short thumb back and forth across the bases of his long fingers. Suddenly he dropped the handful, peered at the side of his index finger and then rubbed it vigorously against his thigh. That clearly brought little or no relief so he rubbed it against the rough bark. I can only guess that he had been bitten by a vicious soldier ant the like of which I too had encountered, when climbing trees to collect leaves for identification. Their heads were rather larger than the rest of their bodies and their massive jaws injected a stinging fluid. I could sympathize with Sam.

Even though it was only ten o'clock I was sweating slowly and every time I stopped to watch Sam, a dense cloud of small *katokali* bees gathered around me and landed to feast on my sweat. In their

131

Beelow sleeping positions

dedicated endeavour they wandered into my ears and up my nostrils; drawing breath to blow them away, I found I had swallowed some others. Sam never seemed to be troubled by *katokali* bees, although I had once seen him impatiently wave away a solitary honey bee. It seemed to be the season for swarming honey bees and almost every day we heard them flying high above the canopy, sounding like an army of minuscule motorcycles, and then saw one of the bright patches of sky way above us turn grey as millions of them passed by in their search for a new roost.

Sam swung on his way picking at fatty grubs with his free arm, when Katy, out of sight but not out of all contact, squealed with a mixture of fear, confusion and anger, and rushed towards Sam. Manai had probably pushed her out of a shared feeding tree again. When the three *beelow* next joined up, they were travelling to a large rambling strangling fig tree laden with bright yellow fruit at the extreme north of their range. It was noticeable that Sam was leading, in contrast with his poor third when Bess was in the group. So far I had only seen Manai lead when she had become scared of me and run off, probably in the direction that Sam and Katy would have taken anyway. So, although Katy was beginning to treat Manai as a proper adult female, the relationship between Sam and Manai still had to be resolved.

I had previously seen Sam leave the strangling fig by two alternative routes; one requiring a huge leap and the other little more than a moderate bound. I watched him climb to the branch I knew he used as a springboard for the leap, and he sat there looking at the space between him and the next tree as though he were measuring it. He bounced on the branch causing it to nod up and down and looked poised to jump. Then he suddenly seemed to change his mind and dropped sedately down to the exit for the safer and easier route, and left the fig that way instead. Even *beelow* appear to get nervous sometimes.

The next food Sam ate was the tip of a vine stem that was swaying around in the light breeze, preparing to fasten onto something sturdy. He snapped off about six inches of the shoot and crunched his way down from the top until it became too tough, rather like eating asparagus. The remainder was discarded. His long arms reached out for another of these leguminous vine tips nearby and he chewed on that too. Meanwhile, Katy had found a column of ants and was busy licking the black crawling mass off her feet, hands and anywhere else they had found before she managed to climb to a bough just above. By noting which hand she used for wiping up ants and for picking fruit, I was able to discover that Katy was noticeably right-handed. Conversely, Sam was left-handed and would position himself on a bough so that he could use his left hand for the delicate jobs, and his right for holding on. It would be fascinating, though perhaps rather esoteric, to know the ratio of left- to right-handed *beelow* in an area. Luckily, however, the time and money required to answer such a problem preclude its ever being resolved.

Manai had lagged a long way behind Sam, probably because she couldn't understand why I was hanging about beneath him. Katy, however, had moved closer to him and, separated by a respectful distance, they entered a gargantuan, broad-limbed dipterocarp tree by the edge of a steep gulley that marked the eastern edge of their range. A vine as thick as an elephant's trunk snaked its way into the high canopy and I could just make out its small leaves and the bright yellow elliptical fruit on which the two *beelow* were feeding. They took fruit in their habitually eclectic fashion, picking from all parts of the crown, never sitting long in one place. Sam finished first and while Katy continued, he ventured to one of the outer branches from which hung

Kasuka ant-plant

134

a host of different-sized *kasuka* ant-plants. These are epiphytes, a little longer than a foot overall, but almost half of this is a grotesque, spiky, black tuberous stem. When I had been scrambling over a new treefall by the Tolailai a couple of months before, I had an opportunity to examine these peculiar plants at close quarters. I could see dozens of small holes in the stems and on yet closer inspection hordes of brownish ants were visible dashing to and fro through these. I cut open one of the stems and found a labyrinth of tunnels inside, some filled with tiny white eggs and others with remains of dead ants and other insects, ant faeces and other bits of frass. The ants and their ant-plants exhibit a form of symbiosis; the ants are provided with security and shelter, and the plant takes up nutrients from the ants' store, allowing it to survive in a harsh but otherwise unexploited niche.

Sam picked seven ant-plant leaves from three different plants and held them in one foot while he swung away, possibly to avoid the biting ants. He selected a small leaf from the bunch and ate it while hanging from one hand, gazing around him. Next he picked a much larger one but ate only the lower part of the leaf-stem before discarding it. It was so crisp, rather like celery, that I swear I could hear him crunching on it. Katy finished eating the vine fruit and came to sample some ant-plant leaves on a bough opposite Sam.

Other gibbon species that have been studied closely, although broadly classed as frugivores, have been found to devote a considerable proportion of each day's feeding time to eating tree and vine leaves. The soft young leaves tend to be more popular than the tougher mature leaves, some species being favoured, some remaining untouched. From Alan's work it was clear that the availability of fruit and tree leaves in our study area was little different from the availability in other areas of gibbon habitat. But the only leaves I ever saw *beelow* eat were from ant-plants, leguminous vines and a few orchids. Why were they apparently passing up the opportunity of an abundant food supply?

It has already been said that there is a low diversity of species on Siberut compared with the mainland and this appears to hold whichever group of animals or plants is considered. Could it not be, then, that with less competition for fruit, the *beelow* don't have to eat leaves? It is, after all, relatively difficult for a simple stomach such as that possessed by man and apes to extract nutrients from uncooked leaves. In fact, when the data I collected were examined closely, they revealed that all the smaller gibbon species spend similar amounts of time each day eating fruit but the *beelow* are exceptional in how little time they spend eating leaves and how much eating insects, thereby getting the protein in their diet from an animal rather than a vegetable source. So, it doesn't seem as though lack of competition from frugivorous species is affecting the *beelow* and it appears that they are actively avoiding leaves, particularly those of trees. The most obvious suggestion to explain this is that the leaves contain something that *beelow* find unpleasant.

This brings me back to the tannins and alkaloids which plants use to defend their edible parts, previously mentioned in connection with the monogamous leaf monkeys. The poor soil and high rainfall on Siberut could necessitate a greater protection of tree leaves there than on the mainland, achieved by producing relatively high concentrations of distasteful chemicals. This could cause the *beelow* (and probably many other potential folivores) to steer clear of them. Orchids and ant-plants rely more on their symbiotic relationship with ants, in their matted roots and swollen stems respectively, for defence and nutrients. Provided *beelow* can cope with the ants their leaves would be palatable. But what about leguminous vine shoots, for they don't have relationships with ants? Well, vine leaves are generally more easily digested than tree leaves; legume leaves are rich in protein; and since the success of any vine depends on its ability to grow as fast as possible up to the light it devotes less of its resources to making toxic chemicals; and the behaviour of the *beelow* seems to substantiate this.

On the way home I stumbled across Attila and his huns, the group of feral pigs that visited our clearing most frequently to rummage through any accessible rubbish, tear down vegetation, to make nests and generally to dig up the garden with their heavy snouts. Attila was a big black and cream porker with the meanest look in his eye and must have weighed half as much again as any of the adult females that roamed with him. He gave a loud snort that scattered the feeding party and I could hear the high-pitched oinking of a new litter.

As I approached the camp clearing I heard Toga start to yell, and by the time I reached the veranda she was crawling clumsily out to meet me. I picked her up and she snuggled into my arms, heaving great sighs of relief. 'I don't know how she knows it's you coming out of the forest,' Jane said from inside the house. 'She never acts the same way when Alan comes home, or when one of the neighbours comes visiting.' Toga wouldn't sit on the floor even while I untied my boots, and looked thoroughly dejected when I put her on the side of the seat next to me. 'What's she done today?' I asked, letting her climb on to my lap again.

'Well, she actually climbed upwards,' Jane replied proudly. 'I'll admit it was only a slight incline but it's better than she's ever done before. We were down by the Tolailai and I hung her on a branch while I was reopening a trap that had closed itself. When I looked up she was nearly out of reach but she soon came scurrying down when she saw me move away. Oh yes, she also made a grab at a little grasshopper, but was so surprised when it tried to escape that she dropped it like a hot brick. Then she caught one of those enormous harvestman spiders and carefully pulled off each of its flailing legs before eating it.'

'Ugh! Perhaps we can stop feeding her milk and banana soon then.'

'I wouldn't bet on it,' Jane said as she brought out the tin of dried milk and improvised baby bottle. 'You'd better feed her now and I'll go and clear up in the kitchen before cooking supper. Fried eel and rice okay? Titikmanai brought us one of those huge green eels in exchange for some beads.'

'Fine.' Bartering food for beads may be corny and outdated but we found that they would buy most of the food and other goods we ever wanted. Although *sikereis* used them in ceremonial head-dresses and aprons, most people had a string or two of beads for everyday wear and were always on the lookout for new shapes and colours, or simply more.

The tall ginger-lily plants at the back of the clearing started swaying and a chorus of snorts and grunts came from the same direction; Attila and his friends were sweeping through. 'The pigs are coming,' I shouted over to Jane, 'make sure there's nothing outside you don't want eaten.' About twenty-five pigs were all around the clearing. From under the house came the unmistakable sound of splintering plastic; one of our specimen bottles must have dropped through the floorboards. It seemed they'd eat anything. That was half the problem because, when we saw them digging up what turf remained in the clearing, we were wont to throw half coconuts at them. Even though I was managing to score direct hits on their muscle-bound bodies, such activity simply brought others running and a scuffle broke out as half a dozen pigs fought over who should have the dry husk. I carried Toga inside and collected one of the sticks we kept by the bed. We used these sticks to poke the pigs to stop them from sleeping just beneath our bed

137

when they had stomach trouble. Back on the veranda, I took aim at the flanks of a belligerent sow and launched my wobbly spear. A split-second later there was an horrendous, blood-curdling scream from the sow and she started racing around the clearing – the 'spear' sticking out of one of her eye sockets. The noise was ghastly but I leapt after her, trying to keep on her blind side until I could get hold of the stick. With a quick twisting tug I pulled it out, then sprinted back to the house and jumped onto the veranda. Revenge was the last thing she was thinking about, however, and I watched her fat behind disappear from the clearing. Luckily, when she returned to the clearing two days later, she was still not vengeful and, although blind in one eye, looked quite healthy. Even so, the episode stopped our more overt efforts at pig clearance and the ground around the house inevitably became like a freshly-ploughed field when dry and like a quagmire when wet.

When the screams of the sow had died away, I sat down with a glass of whisky Jane had poured for me in one hand and a bottle of milk for Toga in the other. She was too tired to insist on banana and was quite content just to suck down some milk. This was the most peaceful time of day. When Toga had finished, I bounced her up and down on my knee for a while to wind her, after which she sat quietly, looking out towards the clearing and the sun setting behind the trees on the far side of the Paitan. What she was actually looking at was anybody's guess. She scratched her nicely rounded, sparsely haired belly, and then her elastic lips started to twitch and her eyes to crease. There soon followed a massively satisfying yawn which ended, as it invariably did, with an involuntary squeak of contentment. She shuffled closer to me and then lay on her back between my legs. She used her long spindly fingers to scratch at the corner of her eye but in fact rubbed everything between her black nose, eyebrow, ear and furry cheek. She yawned again, and reached down with her lanky arms to hold her big toes. Her eyelids started to droop but her dark brown watery eyes kept gazing into mine. She rose to snuggle into my shirt, curled up and almost immediately fell asleep. We were continually struck by the similarities between Toga and human babies; the peaceful trust, the occasional tantrums and tempers, the fickleness, the need for contact and even the erratic rate of development.

Next morning, the first thing I sensed was a clammy little hand across my face and a smooth tongue licking the tip of my nose. I stirred slightly and out of one bleary eye saw a rather muddy Toga nestling next to me. I reached out to put her on the ground but she nibbled my fingers sufficiently hard to wake me up properly. I saw from the alarm clock that the wretched camp poltergeist had switched it off during the night and it was now way past Toga's normal time for breakfast. She had obviously wrestled her way out of the basket in the kitchen and braved the dangers of the clearing to wake us up.

I wandered around between the fruiting trees the *beelow* had been using over the previous few days but the clue to their whereabouts came when I heard the uneven preliminary notes of Manai's song from the north-east. I raced over the small swamp and up the West Ridge towards her just in time to hear the initial rising note of the first great-call. Pathetic; it wobbled upwards, faltered and crashed down again with a cough. After a short pause she uttered a single whoop and then Katy joined her for the second great-call. Katy was very nearly as good as her stepmother although with her small lung capacity she ground to a halt first. They both whooped alternately and with a single false start they produced their best great-call yet. When Bess and Katy had whoooped together between great-calls, it was performed with the panache one would expect from a couple of polished performers. In contrast, this was one of the first times Manai and Katy had sung together and I got the impression that the whoops represented the beats of rival conductors. 'A-one, two, three, four. . . .' 'No, like this, one, two, a-one two three four. . . .' 'What d'you mean? it's one two three, one two three,' and so on. Anyway, they tied themselves into such awful knots that eventually the song bout broke down completely.

Sam meanwhile was nonchalantly scratching the soles of his feet, his armpits and crotch, gazing round at the forested view from his high perch. When Manai and Katy stopped, he produced a single piping note and launched himself along a well-worn aerial pathway next to the Main Trail to the south. As he did so, he passed Manai and Katy but didn't even deign to afford them a cursory glance. In one place, a hundred yards from Summit Two, this pathway crossed a gap in the canopy which was too high for a simple leap even allowing for the long run-up provided by a broad dipterocarp bough on the northern edge. Just to one side of the gap was a very tall nutmeg tree but at the level of the jump, about a hundred feet from the ground, it had no branches or clinging vines. So, to negotiate it, Sam leapt from the dipterocarp onto the trunk of the nutmeg, immediately throwing himself backwards at

139

an angle and twisting his body in the air so that he could catch hold of a branch of the tree on the other side. Whenever he did this, I never failed to experience a great thrill as yet again he performed with such consummate skill. It is interesting, though, to consider that nearly half of a large sample of gibbon skeletons from Thailand showed some healed fractures in their arms and legs, presumably caused by falls.

In the afternoon I came across Jane on her way home after checking the traps. Toga was sitting happily in her arms until she saw me, when she slithered to the ground and hobbled over to me, screaming. The surprised, unbelieving look on Sam's normally impassive face was classic, and he dropped through the canopy like a stone to sit about eight yards away. He uttered a peculiar, ghostlike warble that I only ever heard him make five times, each in a slightly different context but connected by some form of alarm or anger. As he called, so Katy descended to the same four-inch diameter bough on which Sam sat, and squealed unhappily.

Toga appeared completely oblivious of what was going on above her, either because she didn't yet have long-focus vision or because she didn't know what Sam and Katy were. She hooed quietly as I offered

Sam and Katy

her a piece of banana that Jane had passed slowly to me. Katy hooed too, moved towards Sam and, to our utter amazement, they actually embraced each other. It was the first peaceful contact we had ever witnessed between them. Katy then moved back to her original position. I hung Toga on the branches of a small tree next to me and she played happily as long as I stood near her. She pulled off a couple of leaves and dropped them. Then she brachiated to the ends of the branches, the tree bent over as a result causing her to lose her grip and fall to the ground. This made her hoo in distress as she crawled back to me and this time Sam moved next to Katy. They embraced again; a case of unity in adversity. They embraced twice more before Jane took Toga away to save disturbing the wild gibbons any more.

Sam and Katy's new-found unity didn't last long, however, because they slept in separate trees that night, the female being nowhere to be seen. The trees they were using were ones I had recorded as night trees before; some trees were used again and again while other, apparently suitable, ones were avoided. Night trees had to be robust; strong winds sometimes blew at night and a tree that swayed wildly, that is a small one, would be an unlikely choice for sleeping in. Night trees, at least for Sam, had to provide a suitable platform for singing from and would therefore be tall, emerging above the canopy of neighbouring trees. If smaller trees were chosen, his songs aimed at neighbouring and itinerant males would be muffled and their effects reduced. Emergent trees would also be suitable as night trees because, although more exposed to wind and rain, a gibbon would at least not be dripped upon all night after rain had stopped. So, we're left with tall emergent trees as the most likely candidates and it was clear after only a short while that *beelow* do select such trees. Now, from the rate at which Sam used and re-used night trees it is possible to show that he would only ever use fifty trees. However, there are about seven hundred and fifty seemingly suitable night trees in the range; what's wrong with the other seven hundred?

The problem seems to be ants. The *beelow* avoided standing or sitting in columns of ants, they stayed around ant-plants only long enough to pick their leaves and Sam had demonstrated that ant bites caused him discomfort. So with the prospect of spending about fourteen hours out of every twenty-four in a night tree, the *beelow* would try to avoid being bitten. It is not only epiphytes such as ant-plants and orchids that harbour ants, however, for where vines touch tree trunks there are usually ant runways. When I tested my data on a computer, I found that the night trees did, indeed, bear significantly fewer epiphytes and vines than the others. Such delicate ecological relationships are the very essence of tropical rain forest.

Toga's total disregard of Sam and Katy emphasized to us how far she still had to develop before any type of adoption could be contemplated. In an effort to improve matters, and to encourage her to climb higher, I climbed a fruit tree at one end of the clearing with Toga perched on my head. I sat on a branch about twenty feet off the ground and sat her down beside me. Far from leaping gaily around the branches, she buried her face in my shirt and clung on in sheer terror. What rotten luck to have an acrophobic gibbon! Now and again she'd twist her head very slowly and then, equally slowly, peer down. But as soon as she saw that the ground had come no closer, she buried her face again and whimpered. Toga still had a long way to go.

Chapter Eight
Sikereis, Shields and Songs

While Aman Bulit was living on the Paitan building a new *sapo* downstream from us, I asked him to take a day off from lugging tree trunks out of the forest for the corner posts and help me collect plant specimens. When Alan left for England he took a dozen bundles containing hundreds of leaves soaked in alcohol for identification at an herbarium, but my *beelow* were still eating fruit from uncollected tree species and the work had to continue. The ease with which Aman Bulit pulled himself up the vines and trees with his muscular arms was amazing to watch and he showed considerable staying power. When he disturbed vines against trees, infuriated ants living between the two crawled all over him. He allowed himself only a few quiet *'teelays'* and once on the ground again I found him studded with soldier ants, their heads a quarter inch across and heavy jaws sunk deep into his skin. Siberut people have a higher 'grumble threshold' than Europeans, it seems, and will for instance walk on for half an hour to a convenient stopping place before cutting painful *ariribuk* spines from the soles of their feet with their *parang*.

We were walking along the Fig Trail towards the next tree from which I wanted leaves when I noticed a peculiar tripod of leaves and branches at the side of the path. It was obviously man-made, standing about a yard high and consisting of leaves from three or four types of shrubs and palms bound with strips of bright red cloth to the tip of a large palm leaf. Aman Bulit came up behind me and, anticipating that I would, as always, want an explanation, looked around for a convenient but not too rotten log to sit on.

'*Oto, sikine*, it is like this,' he began, 'these leaves are a present to the forest. I told you once before that everything has a soul and everything that has a soul has *bajo*.' I nodded attentively. 'Well, even though *sikereis* can speak with the souls of our ancestors, speaking to the rest of the invisible world is very difficult. Now, luckily there are certain objects, usually plants, that can act as mediators. *Sikereis* know that the souls of certain mediators are prepared to tell the other

souls of our wishes. And these mediators can encourage the granting
of a wish. I think those leaves were placed over there by Aman
Mynoan in an attempt to get back a small herd of his pigs that have not
returned to his *sapo* for over a week to eat their sago. He knows he can
get them back as long as he uses the correct combination of leaves.'

'But why put them in the forest? Surely he could ask the mediators'
souls to go from his *sapo*?'

'Ah, that is what I meant when I said the bundle was like a present.
You see, not only does everything we can see have a soul, but there are
also souls of things we cannot see. There are *tai ka koat* spirits in the
sea, *tai ka manua* spirits in the air, *tai ka baga* spirits in the world
beneath the earth, and *tai ka leleu* spirits in the forest whose chickens
are the monkeys and whose pigs are the deer. These leaves are both a
beautiful gift for *tai ka leleu* and a way of asking them to return his
pigs. You see?'

'Yes, I think so.' The concept of forest spirits at least was very easy
to understand because it was impossible not to sense a peculiar 'presence' whilst under the lofty, imposing canopy.

'Pigs and chickens are also mediators,' Aman Bulit continued, 'and
they can give some clue as to whether the *sikerei*'s request has been
granted.'

'How's that?' I asked him.

'After the *sikerei* has appealed to the soul of the animal to act on his
behalf, the animal is killed and its heart and the skin between the
intestines are examined. I cannot read the signs, but a *sikerei* can tell
from the blood vessels whether his request has been successful.'

Back in camp later that afternoon, Jane and I began planning our
final long survey and we took the opportunity of seeking Aman Bulit's
advice on feasible routes through some of the southern river basins. He
had come to know the area very well when he lived with the Sakuddei
clan for two years with Reimar Schefold, and it seemed obvious to ask
if he'd guide us. He took a little time to answer – it would have been
rude had he not – and we were tremendously pleased and grateful
when he finally agreed. We planned to travel clockwise through the
basins of Silaoinan, Sarareiket, Sagulubbek, Paipajet and Simatalu.
Toga would have to come with us, so that our by-now-minimal survey
kit of one two-foot-six camping mat, one mosquito net, two *sarongs*,
several packets of dehydrated food and half a dozen slabs of tobacco
would have to include her sleeping basket, a bunch of bananas, bottle
and dried milk. We didn't anticipate any trouble with her just so long

as we were able to buy bananas at regular intervals.

We left Sirisurak a week later and spent the first night in Katinambut, the most northerly village in Silaoinan. It was a wonderful collection of large *umas* quite unlike the small houses found in most villages. We stayed with a man who, when a child, had been 'adopted' by Aman Bulit's parents because he'd looked so much like their first-born son who had died while still young. Shortly after we arrived we ascertained that a 'party' would be held that evening in one of the adjacent houses as part of the induction of a new *sikerei* and we asked whether we might attend for some of its duration. About four hours later news came through that we would be welcome and so, leaving Toga sleeping soundly in her basket and pausing only to tear off half a slab of tobacco, we hurried after our host who lit the way with a flaming bamboo torch.

Having attended a ceremony for the dedication of the new Sagaragara *uma* in the Paitan headwaters we knew something of what to expect, and this was very much a 'modern' village affair with the majority of onlookers wearing western-style clothes but the *sikereis* and their wives dressed traditionally, their skins tinted yellow from having rubbed on tumeric. The three *sikereis*, one pupil and two teachers, were dressed identically and wore smart, thick, brown, bark loincloths, portions of which were wound round with red and yellow strips of cotton material. In front of the loincloths hung loose, trapezoid barkcloth aprons dyed red, black and white and held in shape by a rod along the base. Around their necks they wore red, blue, orange and white glass beads as well as a panel of beaded designs, chicken feathers and amulets. Small glass beads were also set in patterns on headbands and on armbands of barkcloth that were worn just above the elbow. In addition, the *sikereis'* long hair was bound up with red cloth in a horizontal bundle containing helpful mediator objects. Leaves from mediator plants were tucked into the back of the loincloth like a large green bustle, and into the headbands and armbands.

The *sikereis's* wives sat in a straight row in front of the other women and wore bright scarlet skirts with strips of yellow and green cloth sewn on in bands. Around their waists and necks they carried countless rows of red, black and light blue beads, some threaded on thin strings of spun bark and others on lengths of stiff wire that held the necklace in an exact circle rather than following the contours of the wearer. Most unexpected of all were the wives' magnificent headdresses or *teteku*. They were similar in some respects to the mens' headbands but built up on a base of rattan strips was a halo of mainly white feathers looking somewhat like the archetypal head-dress of a North American Indian chief. Among the white feathers were tight bundles of variegated feathers, discs of mother-of-pearl from nauptilus shells, and red and white flowers from hibiscus and lilies.

We were extremely relieved that our arrival at the ceremony caused almost no fuss and for the first fifteen minutes most people sat smoking around the edge of the veranda; the three resplendent *sikereis* sat in the centre. A couple of young men passed to and from the hearth inside the house carrying *kajeuma* drums. These were a yard long and made from the hollowed trunk of a *pola* palm with a python skin stretched across one end. They had to be warmed so that the tension on the skin was correct and the *rimata*, the 'master of ceremonies', tested this each time a youth brought one out. He was prevented by taboos from warming the drums himself. Eventually the *kajeuma* sounded to the satisfaction of the *rimata* and the youths sat cross-legged to one side of the veranda, rested the drums over their left thighs and with their right hands started to beat out notes to an even rhythm in unison. A third youth behind them picked up a small cooking pot and used a monkey femur to beat out the same simple rhythm but lagging about a quarter of a beat behind the others. He concentrated very hard to get his deceptively difficult part correct but kept slipping into the drums' rhythm. In exasperation the *rimata* came to sit next to him for a while and literally held his hand.

After a minute or so of competent drumming, the *sikereis* rose to stand in a circle and the chattering crowd quietened down and focused their attention on the dancers. Each pulled two leaves of a ginger-lily from the 'bush' in his loincloth and held these delicately between index and middle fingers with the little fingers pointing up and out as one might at a refined tea party. They bent their bodies slightly forward, held the leaves in front of them, elbows away from their sides, and when the drummers beat harder, the oldest *sikerei* started to sing a mournful, slow verse and the other two joined him in unison.

Suddenly they stopped singing and the house shook dangerously as a sound like rhythmic gunfire began. The *sikereis* were stamping their horny heels with small but very strong movements. It was then that I realized why the floor they were standing on was made of rough hewn, loose planks instead of the usual *ariribuk* trunks. Apart from the planks' strength they made an amazing din and everybody's feet involuntarily mimicked the beat. The basic theme played by the feet was: ⌐|⌐⌐|⌐|⌐⌐|⌐ but they ventured through innumerable syncopating variations of this. Then, when we least expected it, the stamping stopped, the *sikereis* moved their garlanded arms up and down with graceful, flowing movements, moved forwards to the position that had been held by the *sikerei* in front of them and they began singing again. In due course the stamping started again and so the dance continued.

When it was over, the *sikereis* broke ranks and leant against walls or supports while women fanned them with large sago leaf-sheaths. The *sikereis* also fanned themselves with their aprons and consulted with each other. Still seated, they sang a more melodious song accompanied only by the chirping and rasping of frogs from the black night outside. This song was slow and solemn, and they adopted the peculiar falsetto register used when singing many of the traditional songs. It was a beautifully ethereal song about the tree-swift and the notes rose and glided, just like the graceful bird they sang about. I was surprised I couldn't catch more than a handful of words, but later Aman Bulit explained that *sikereis* sing songs and chant invocations to the souls in an artistic and refined bardic language that is the same all over Siberut.

Stamping dances followed seated songs followed stamping dances until some hours later one of the *sikereis'* wives stood up in the middle of a dance, eyes shut tight, and joined the circle. She was in a partial trance and acted like a blind woman colliding with the *sikereis* who were 'flying' round their circle. At the end of the dance the *sikereis* slumped exhausted to one side of the 'dance-floor' but the woman kept trotting round and round, arms out, faster and faster, more and more excitedly, making the planks jostle and knock together. At length, her husband started to chase her; sweat began to run down their bodies and the movements of their limbs became looser. Near the entrance to the house, the oldest *sikerei* was seated, eyelids shut, and beginning to sing as he swayed to and fro seemingly oblivious of the exhausting scene next to him. Suddenly, after a quarter of an hour of crashing feet and beating drums, the woman collapsed in a heap and was pulled to one side. The old singing *sikerei* interspersed his sung phrases with chilling wails and his arms snaked around his body, conceivably independent of each other and of him. Soon his whole body writhed and twisted; his arms now above, now in front, now behind. With no warning, all the strings in his body seemed to be pulled tight, he sprang backwards into the arms of a youth who, although expecting the spasm, obviously didn't reckon for the force of it. The *sikerei*'s body was rigid, arms by his sides and legs straight out. This tranced state is taken as proof by the people that a *sikerei* is associating with the spirits and ancestors' souls and during it he may receive advice on conduct or new taboos or a 'recipe' for a new healing potion. Such instructions are generally given by the soul of the particular deceased male relative who appeared to him during his very first trance and who remains as a life-long adviser.

The *sikerei* who had chased his wife now put a small bell into the clenched fist of the tranced man and sprinkled water from a coconut shell onto his hair using leaves of mediator plants. Some minutes later the first trembled perceptibly and the bell sounded softly; slowly the ringing became more and more certain and the *sikerei* sat up stiffly. His confused eyes opened reluctantly and all knew he had returned from his turbulent, distant spiritual journey. He turned his attention to the wife who still lay in a trance and he summoned some youths from the inner room. They entered, each carrying a bundle of smouldering kindling and handed these to the two experienced *sikereis*. They in their turn each held one of the woman's arms and then stroked the smoking wood up and down the inside of her forearms. The two of them then sang over her body and rang bells by her ears. At last she came to and there wasn't a mark, not even a reddening, on her arms.

The three *sikereis* sat down to rest together and sang some more haunting songs about the forest and its inhabitants. Towards the end of the evening the dancing took on a different style. The story was enacted of a confrontation between a *joja* and two *beelow*. The new *sikerei* played the *joja* which was feeding peacefully in a tree when along came two *beelow* who taunted and goaded him until he left the tree. They all chased around the veranda, climbing up posts and swinging on beams, and gave excellent imitations of the appropriate calls. The audience (the mortal one at least) enjoyed the dance enormously and at times the laughter and screams of delight almost drowned the rattling planks.

We left Silaoinan laden with food for the road. Toga had decided she wanted to walk so, as a compromise, Jane carried her by one arm like Christopher Robin carrying Pooh. The first hour was atrocious, for we were constantly slipping off submerged logs beneath seemingly endless sago swamps, but then we started climbing. At the top of the divide between Silaoinan and Sarareiket we took a well-earned rest in a small glade where the remains of sago sticks on the ground attested to the need by all to stop after the steep ascent. A rest was always the cue for Toga to work off some of her frustration at being carried, by rushing around trying to initiate games of tag and Aman Bulit was always her best bet.

'What makes a man become a *sikerei*?' I asked Aman Bulit in a moment of calm. 'I mean, why do some men become *sikereis*, some *rimatas* and others nothing at all?'

'It is simply a matter of choice,' he said with Toga lying panting on his chest. 'You do not gain anything for yourself by being a *sikerei* because, although it makes you someone rather special, there are many extra taboos you must follow.'

'I thought *sikereis* received payment of food for their services,' Jane said.

'That is true,' he told her, 'but he has to take the food back to his *uma* and share it among his clan.'

'What about *sikereis*' wives, are they trained too?' she asked.

'No, but at the start of her husband's training they must live alone together in the forest and she too must accept many new taboos. She does not learn healing, songs or dances from the *paumat*, the experienced *sikerei* chosen to give instruction, but she will sometimes help when her husband appeals to the souls for help or advice.' He paused for a moment to give Toga to Jane. 'Of course,' he continued, 'it was a *sikerei* who made the Mentawai Islands.'

'How do you mean?' Jane asked while feeding Toga on one of our replenished store of bananas.

'Well, it was Pageta Sabbau, the first *sikerei*; a man with many magical powers and the man who gave each river basin its own language. He lived on Siberut when it was the only island in the great ocean. One day his nephew challenged him to make the south of Siberut fall away into the sea. Pageta Sabbau took his bell and climbed to the top of a hill from where he could see all the land to the south. He rang the bell and sang magical songs to the souls, and slowly three pieces of land drifted away. A fourth piece was just about to tear off when the clapper of his bell broke and the land stopped moving. That is how the two Pagai Islands and Sipora were formed. The large peninsula in the south-east is the land that nearly became an island.'

'How was Siberut formed?' I asked, prolonging our rest just a little longer.

'That is not known,' he said flatly, but the omission in the folklore didn't worry him. 'It is known, however,' he went on, changing the subject, 'how a *sikerei* domesticated the first chickens and pigs.'

'Go on,' I urged, as he was clearly fishing for an invitation to tell another story before we continued on our way. Toga, full of banana and feeling contented, had crawled over to me and was now asleep in my arms, stirring every now and then to scratch or fan away an over-curious mosquito with her long fingers.

'The second *sikerei* was a man named Maliggai,' he began again, 'a member of the Sapojai clan – the clan Ohn belongs to. When Maliggai was just one month old his mother died and he was laid to rest with her in a wooden coffin and taken to a distant hill for there was no one in his clan who could take care of him. Three days later, Oakoak, an old man from the Salabekeu clan passed by and heard Maliggai crying. He wanted to adopt Maliggai for he had no children of his own, so he asked the Sapojai clan for their permission. He took Maliggai home to his wife and she washed him and cared for him. Maliggai was not very old when he began to talk – but he spoke in a strange language, the *sikerei* language, that his foster parents could not understand.

'Then, when Maliggai was about ten years old, he told Oakoak that he was going to become a *sikerei*. He told all his friends to build a pig pen below the house but none of them knew what a pig was. "*Tak leu anai kudnia*, never mind, do as I say," Maliggai said, "just collect heavy stakes, make a fence below the house and put in some split sago trunks. Make a sliding door at one end too." Then Maliggai told everyone to make chicken baskets but none of them knew what a chicken was. "*Tak leu anai kudnia*, never mind, do as I say," Malaggai said, "just collect *pelege* rattan, split it and weave an open basket two hand spans across. Fasten these below the eaves."

'When all these things were ready, Maliggai stood at the front of the house and called "*Ki-or, ki-or*". Slowly, a herd of pigs came out of the forest and the people were frightened for they had never seen pigs before. "Do not worry," Maliggai said to his friends, and the pigs came in to the pen and ate the sago. He lowered the sliding door and then called out "*Ei-a, ei'a*". Slowly, a group of chickens came out of the forest and the people were frightened. "Do not worry," Maliggai said to his friends, and they watched as the chickens entered the baskets and began to lay eggs. Maliggai's friends asked him, "What are these animals for?" and he told them they were for food at ceremonies. So, they all began to make beads and new loincloths and

150

to collect red dye from the mangrove trees. Then they celebrated by eating some of the pigs and chickens. During the ceremony Maliggai healed sick people and as his powers became better known so he was called to all parts of Siberut to give medicine and to train *sikereis*. That is just one of the stories about Maliggai, the second *sikerei*.'

'Maybe,' Jane chuckled as she stood up, hooking her arms under the straps of her rucksack and waking Toga, 'but we can't listen to stories all day. We have work to do.'

The *uma* of the Sabulau clan in the headwaters of the Dereiket river was the first fully functional *uma* either Jane or I had seen. One hundred feet long, it stood imposingly at the top of the river bank behind three tall bamboo poles from the top of which hung carved wooden birds dating from the dedication of the *uma* a few years previously. We climbed up the thirty steps cut into two long tree trunks leading to the entrance platform and above this stood an archway crowned with three more carved and painted birds. It was here that we were greeted by the old *rimata*, Aman Palibatti, and he

Door handle

151

beckoned us inside. We had intended merely to drop in, but there was a heavy storm brewing and from the moment we tripped over the carved step into the veranda we knew we were entering a rare oasis of Siberut art.

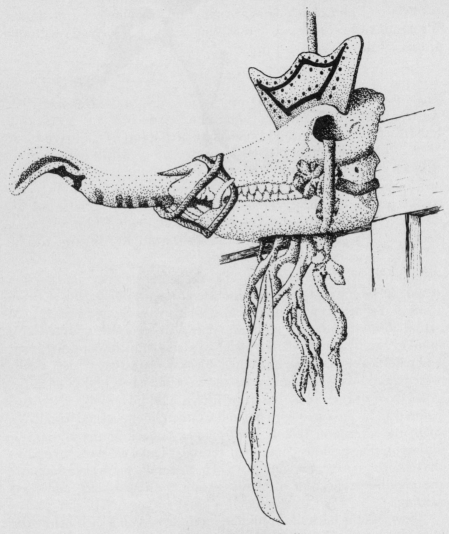

Decorated pig's skull

Above and inside the entrance to the veranda were tied skulls of about thirty feral pigs that had been sacrificed for clan ceremonies over the years, and in the centre was a carved pig's head painted and

Carved pig

decorated with a slender green cock's tail feather. In the middle room where the young and older children slept was the *baibai*, the beam in front of the large central fireplace on which were hung the decorated skulls of wild animals – monkeys, deer and boar. We counted 250 monkey skulls, almost all of which had plaited bunches of dried grass and palm leaves hanging from the lower jaws; the twenty deer skulls were painted with lines and dots and two had carved birds on short poles fixed to the skulls' base; the fifteen wild boar skulls were also painted and were fitted with painted wooden 'tongues' and 'ears'. We spent the afternoon talking with people, collecting animal names, sitting on the spacious communal veranda. Later we were invited to the back room to eat and on the way I accidentally hit my head against the ghoulish collection of skulls setting the loose teeth and jaws rattling.

Towards the back of the middle room, stored on a shelf below the eaves, was a *koraibi* or shield about three and a half feet tall with a handle cut into the wood that had been covered with half a coconut shell as a finger guard. Aman Palibatti told us that they were used until about fifty years ago by men in the front of a head-hunting band to protect the archers behind them who needed both hands to aim and

Koraibi

shoot. This one had been kept because at the top was a drawing of the now-dead owner's hand. This representation of a deceased clan member was developed on the *kirekat*. This was a plank on the back wall on which were painted quarter moons and a man's hands and feet. Feet are the most common symbols used on *kirekats* because people can be recognized from their footprints alone, so often seen in the ubiquitous mud.

Just behind the low door to the back rooms hung the *bakkat katsaila*, a bundle of particularly effective mediator plants and the *uma*'s main fetish, but above it was a large *jaraik*. This is an

Kirekat

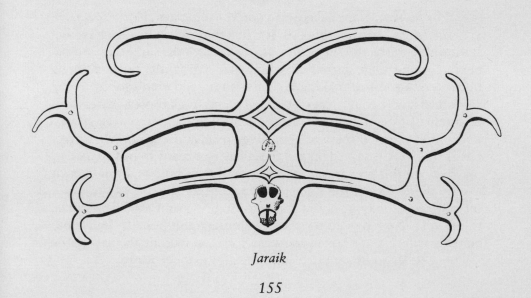

Jaraik

elaborately carved board with a *bokkoi* skull mounted on it. 'The *jaraik* is very important to an *uma*,' Aman Bulit told us, 'because it attracts good forces. It is very difficult to make a *jaraik* and that is why, especially today, there are so few on Siberut. When this *uma* was completed, many were invited to come here to celebrate. When they were all gathered together they went hunting but it was different from all other hunts because they sought only a large male *bokkoi*.' He confirmed this with Aman Palibatti.

Jaraik design

'Why *bokkoi*?' Jane asked; 'is a *bokkoi* soul particularly special?'
· '*Tak taagai*, I do not think so. It is just that a male *bokkoi* has a big skull with large teeth. The people ate the *bokkoi* the same day and the next morning they started chiselling out the *jaraik* from a single buttress of a *gite* tree making it similar to the old *uma*'s *jaraik*. They smoothed it with rough leaves, painted it and stuck those smaller discs of mother-of-pearl onto it. Pieces of mediator objects were placed under that larger piece and that gives protection from evil forces. Lastly the skull was fixed on and the *jaraik* was ready to be fastened to the wall. All this preparation takes several days but, at the end, each man of the *uma* sacrifices one of his chickens for the soul of the *jaraik* for it attracts good forces and it is asked to bring souls of wild animals to the *uma*. Next day all the men go hunting and hope to find those animals whose souls are in the *uma* for they will let themselves be killed in order to be reunited with their souls once more.'

Visitors to an *uma* or *sapo* generally sleep on the veranda but we were allowed to sleep in the middle room. To save weight on surveys Jane and I carried only one sleeping mat and mosquito net between us and this had caused some comments from people in the past. Sexes usually slept apart in an *uma* and making love there is prohibited by taboos. When Aman Palibatti and other clan members saw our sleeping arrangements they were rather anxious but we assured them that our only intention was to sleep in preparation for the next day's walk.

When we reached Bat D'erat, the first river in the Sagulubbek basin, we discovered why Aman Palibatti had advised us against travelling during even a minor flood. Part of the 'path' followed a winding narrow gorge the sheer sides of which bore only those few plants whose roots were tenacious enough to provide a firm anchorage. By craning our necks we could see the forest way above us; the sounds of water dripping from the rain-splattered vegetation, together with our wading through the deep water, echoed about us creating an uncanny atmosphere.

Shortly after negotiating the gorge, Aman Palibatti and his son who were guiding us found someone's dugout tied up at the bank. I heard no discussion on the subject of whether or not we should take it, but Jane and I were grateful enough to get in. Whoever had left it there so conveniently hadn't left any paddles, so our guides had to make do with long bamboo poles which, where the river was deep, were little use. There was much '*teelay*-ing' when we turned a corner and finished up pointing the way we'd come – just as Jane and I had done (with paddles) when we were learning to control a dugout.

A short way further on, Aman Palibatti, his Edwardian style hair now in disarray, asked us three to get out and walk while he and his son took the dugout through the *tattanan*. Only after we'd climbed to the top of a spur did we find out what this was. Below us we saw another narrow gorge but with the quantity of water above it, it was a boiling, foaming maelstrom, made worse by a wicked right- and then left-hand hairpin bend. The guides edged gingerly away from the calm bank where we had left them, and knelt in the bottom of the dugout with their knees pressing against the sides. Both luck and judgement helped them as they ducked and bobbed through the gap but before we could resume our journey there was a lot of baling to do. As we continued downstream Aman Palibatti told us about all the people who had perished because they had not reckoned with the force of water in the *tattanan*.

It came on to rain heavily, the whole surface of the river now pitted by the large drops, and baling had to be quickened as water was coming in from all angles. A creeping chill rose from our numb buttocks as we sat still in the dugout, the only warm patch for me being where Toga was pressing herself against my chest in an attempt to keep at least some of her body dry. Our own discomfort was relieved, however, by the majestic scenery of tall trees that leant out from the sloping banks and touched branch tips way above us. Vines bound the two sides together and some of those rooted on the left of us were dropping fruit into the quickening flow on our right. When the rain eventually stopped, birds flew between fruit trees under the corridor of green and we chased spectacularly colourful, large kingfishers for hundreds of yards.

In the mid-afternoon the forest became more patchy and we crossed between the two main landforms on Siberut – the sloping hills and the flat flood plains. A type of wild sugar-cane fringed the unstable banks and behind them we could see banana fields. Aman Bulit was becoming more and more excited because it had been three years since he had visited the region and he began shouting a piercing note in a falsetto voice as we rounded every bend. Then, at last, there was a reply. Then we saw smoke, then a small thatched roof and then a muscular *sikerei* standing at the top of the bank, his pale loincloth and very long hair billowing slightly in the breeze.

'*Aleh, anai sibulau*, we have brought some "white" ones to you,' Aman Palibatti called from the back of the dugout. The *sikerei*, Aman Uisak of the Sakuddei came down to the river's edge and hugged Aman Bulit. He then greeted Jane and me in a wonderfully warm and genuine manner, and a rapport emerged that was quite unlike any other temporary relationship we had made on a survey. The fact that I knew Reimar Schefold was the best introduction Aman Bulit could have given, and we were immediately invited into his *sapo*. As usual on entering an unfamiliar house, Toga was subdued, not biting or screaming, and very cute. She allowed Aman Uisak to pick her up, stroke her and he kept telling us how the fur, limbs and digits of *beelow* differed from the Siberut monkeys. Toga was a trifle bemused as her legs and arms were lifted and her unexpected malleability made her a great hit. 'We shall have to call you Aman Beelow,' he said to me, 'and Bai Beelow,' to Jane. The names stuck.

Simakobu

When Aman Uisak had built his *sapo* he had chiselled the planks at the back of the veranda so as to leave large carvings of animals raised above them. There were *beelow*, *simakobu* and *joja* as well as the woolly-necked storks and hornbills. There were also monkey skulls all of which had some fresh and colourful leaves hanging from them. Around his clearing a couple of healthy-looking pigs were chewing at a section of sago trunks and a dozen brown hens scratched in the dirt. Bai Uisak, a small, amply rounded and cheerful woman, cooked us sago and a chicken stew, and after we had all sat down around the food Aman Uisak said grace, not in the more usual Christian style but in a traditional form.

> 'Oh, ancient ancestors, we present you with food as we ask
> you to cast aside our sicknesses, our anger and those of our
> children and friends. Do not go away to the lightening; it
> scares us – here is your food and our food, your fish and
> our fish. Here is a meal. Come, enter, let us all eat in
> this house together in peace – tulutulutulutulutulu!'

It would be wrong to say that Aman Uisak and the others of the Sakuddei clan live wholly traditionally because styles evolve continually and enterprising traders began to import metal blades to replace stone tools as long as several hundred years ago. It would be correct, however, to say that the remote Sakuddei clan live closer to the 'old-fashioned' style than almost any other group. We began to think that Cambridge would never miss us if we built a *sapo* like this near a river, used our savings to buy a small herd of pigs, some durian, coconut and other fruit trees, and got practising with a fishing net and a bow and arrow. But could we stick it? After more than two decades of conditioning, could we stand the high infant mortality, the monitors and pythons eating our cosseted stock, the cobras eating our chicken's eggs, the floods, the *ariribuk* spines in our unprotected feet?

The Sakuddei can do it happily, almost joyously, but those in the past who couldn't have fallen by the wayside. We decided just to be very grateful that we had the rare opportunity to experience life in this part of Siberut.

Aman Bulit erected his mosquito net next to ours when the time came to sleep. It was 31 December and we wished him a Happy New Year. He was the only other person for miles who knew what a year was.

We were woken before dawn by the sound of Aman and Bai Uisak grating coconuts and cooking sago respectively. Aman Uisak had grated five juicy coconuts the previous night and now he was busily grating another five, all for the dozens and dozens of chickens that were now milling around the *sapo* in anticipation. We began to appreciate the enormous wealth of the Sakuddei. Some of Aman Uisak's pigs were attracted by the flapping chickens and about thirty of them, mainly black but of all sizes, were soon snorting, snuffling and squealing on the ground below us. Toga leapt excitedly over the floor sticking her bottom in the air as she peered at the herd through the gaps between the floorboards.

An hour's dugout ride away the well-settled, mottled moss-and-lichen-covered thatch of the Sakuddei's *uma* was camouflaged against the forested hill behind it. The whole *uma* was showing its age, almost thirty years, but what it lacked in right angles it made up for in character and a sense of history. We were given an effusive welcome by Topoi Oggok, the clan's *rimata* who was not in the least surprised to see us. He had killed a chicken that morning and the blood vessels around the intestine clearly showed a visit from new and old friends but he admitted he couldn't quite tell who we'd be.

Topoi Oggok was an amazingly vociferous, marvellously spritely old man with surprisingly taut muscles, heavily creased skin and only a few rather brown teeth. It was soon apparent that his one failing as a raconteur was not his hopelessly hoarse voice but that he laughed endlessly at his own jokes. We told him why we were living on Siberut and he loved the idea of calling us Aman and Bai Beelow. He said in a loud aside to me, 'It is lucky that you brought Bai Beelow with you. If you had not, I think that when you returned to your *sapo* you would have found she had run into the forest to live in a tree with a *loga*!' I thought he was going to choke himself in his spasm of laughter.

160

Our tobacco was being put to good use. In contrast to the method of making long, thin cigarettes with palm leaves that we were used to, many of the people in the south of Siberut use spiralled strips of young banana leaves. This produces a chubby cigarette not dissimilar to the stub of a fat cigar. Topoi Oggok spat through the floor and suddenly remembered something. He recalled, with actions, how the Granada TV film crew who stayed at the Sakuddei *uma* with Reimar Schefold for three months had blown their noses into white cloths; he presumed they did this to take the contents back to their village. Dirty habit!

As more bananas were brought out by Topoi Oggok's wizened wife, so another woman came to join us. She was Topoi Oggok's daughter, Bali Kerei, who went directly to sit by Jane. As with all the clan members, Bali Kerei understood our difficulty with their language and spoke simply, slowly and clearly and we surprised ourselves how much we could communicate; we learnt more of the southern Siberut language during our short time in Sagulubbek than we ever did at Paitan. We estimate that Bali Kerei was nearing thirty and it seemed strange that while very attractive she was still single at an age when most girls would have been married for at least ten years. Once, when they were alone, Bali Kerei told Jane how much she'd like to be married. Jane asked her why she wasn't, to which Bali Kerei replied simply that her bride price was too high. The Sakuddei clan was so rich that the men had no difficulty paying the necessary pigs and fruit trees for any bride they wanted. But none of the men in the area could afford to marry her, and as time passed so her skills and usefulness to the clan increased and her price rose commensurately, compounding her difficulty.

In the afternoon Topoi Oggok took me with him to feed the chickens that did not live at the *uma* but were scattered between small shelters. For half an hour beforehand he had been mixing quantities of grated sago pith and coconut and it was apparent after the first stop that this combination works wonders. There were literally hundreds of chickens and all as tame as one could wish. Topoi Oggok used Maliggai's traditional chicken call, '*ei-a, ei-a*', to gather them around him and he spoke soothingly to them, stroked them and a more contented assembly of chickens would have been hard to find. As we walked to the next chicken hut, Topoi Oggok caught hold of the back of my trunks and yanked them up. '*Tak maeruk*, that is not good,' he said. Then he pointed at the front of my trunks and repeated, '*Tak*

161

maeruk, that is not good.' Personally, I thought I was dressed very adequately, particularly since Topoi Oggok would have been arrested in Europe for wearing his skimpy loincloth, but as soon as we returned I changed into a pair of shorts that flattened my curves and hid the top of my buttocks. Similarly, Jane changed out of her shorts and into a long *sarong*.

Back at the *uma* Topoi Oggok sat between us and everyone started talking about the strange differences between cultures in what one can, and cannot, reveal. Jane was told it was good her thighs were now covered but she didn't have to bother with her chest. She responded by telling people in a mishmash of languages about the shifting prohibited parts of the human body in Britain over the last few hundred years. Stories of Victorians who even covered over the legs of tables to save embarrassment were greeted with choruses of incredulous laughter.

Before supper Aman Duman Kerei, a sibling of Bali Kerei and Aman Uisak, arrived at the *uma*. He was less well built than Aman Uisak and had his hair pulled back tightly into a bun instead of wearing a long pony-tail, but he was identical in the warmth of his welcome. Later, everybody sat on the veranda and listened to songs and stories. Songs are being composed every day on Siberut, and Aman Bulit had often sung us some of his, the tunes of which were influenced by the western songs that blared from the few radios in the villages. Not so the songs of the Sakuddei; they were of quite a different genre and similar to those we had heard during the ceremony at Katinambut. Aman Uisak sang about *beelow*, orioles, fish, bats, butterflies and about the sounds of the forest; Aman Duman Kerei told a story about the Earthquake spirit; and Bali Kerei sang her own song about the two buffalo that had been presented to the clan by the previous governor of West Sumatra. It was an indication of the *uma*'s remoteness that the word they used in songs and stories for other Indonesians was '*sasareu*' or 'the people from the far away land'.

The next morning Aman Bulit, Jane and I were up early into the hills to listen for primate calls. When we returned to the *uma* at midday, Bali Kerei came out onto the veranda with a large platter or *lulak* of food. The sago, coconut and *subbeh* taro balls were predictable but not so the massive *tamara*-like grubs at one end. 'They are *tutube*,' Aman Bulit told us, looking at our questioning faces. 'People cut around the bark of a large tree such as *koka*, and after maybe six months fell it and chop out the *tutube*.' We discovered later that the grubs were the larvae of longhorned beetles that were themselves almost four inches long. By now fairly catholic in our tastes, Jane and I tucked into the grubs and agreed with Topoi Oggok that they were even more delicious than the *tamara* from sago.

After lunch, Bali Kerei asked Jane to help her make roof pieces for the re-thatching of the inner section of the *uma*. Each roof piece comprises about thirty sago leaflets bent across a piece of split bamboo and sewn together with rattan 'thread'. As Jane found out, however, there was considerably more to it than met the eye. Each leaflet was bent, not in half, but about three-fifths of the way along from the tip; there is a precise point relative to the veins at which the rattan enters and leaves each leaflet; and each stitch has to be slightly out of true so that, somehow, when the rattan is pulled tight it becomes straight. Bali Kerei was a thoughtful, patient teacher and eventually Jane felt she was actually contributing rather than being a liability. During the course of the afternoon, Jane and Bali Kerei also agreed to become '*siripoks*', a life-long relationship best translated as 'blood sisters' (although blood is not mingled). It recognized the close tie of friendship that was developing, despite the language problems, and was marked by an exchange of gifts.

Two days later after more and longer forest surveys Aman Duman Kerei came to find us just after dawn to ask if we'd like to go with him, his son, and Bali Kerei, to castrate a pig. Indeed, we would. It took about twenty minutes to reach Aman Duman Kerei's *sapo* by dugout and when we arrived the eighteen-month-old pig that had been confined in the small trap the night before started to squeal his head off as if he knew what to expect. The trap had walls about seven feet tall, a design born of the experience of how high pigs can jump when they try. To extract the beast, Duman Kerei lowered a long, hefty noose of rattan cane, similar to the ones used for catching chickens, and managed to hook it round a hind leg. He and his father hauled up the indignant, flailing pig, grabbed hold of its muddy, slippery limbs as best they could and manhandled it onto the floor of the *sapo*. They bound its legs and jaws but Duman Kerei still had to kneel with his full weight to ensure that in the struggling Aman Duman Kerei didn't cut off anything he shouldn't. The operation was swift, simple and helped along by Bali Kerei making rather apposite jokes. The pig didn't race off into the forest or turn on its tormentors as I'd expected; instead it ambled peacefully under the *sapo*, accepting hand-proffered bananas.

Bali Kerei had been boiling a couple of eggs and after everyone had washed their hands they were offered to Jane and me as snacks. 'Why don't we share them,' Jane suggested.

'*Tak leu*, never mind,' Aman Duman Kerei excused himself, 'it is taboo for us to eat eggs during the day we castrate a pig.' Traditionally, the Siberut people see all their activities as a disruption of the harmony in which the rest of nature stands. Since all things in the physical and spiritual worlds are in a form of contact that knows no boundaries, any disturbance can potentially alter delicate balances between forces and precipitate unimagined effects. So, when a man wants to change something to suit himself, he has to consider his position within the multidimensional matrix of interplaying forces. Mediator objects help smooth out any unexpected changes but these alone are insufficient. It is necessary, in addition, to compensate for the imposed change by doing something in return. This, then, is the root of the taboo system in which abstinence from some activity is required. Sometimes the taboos can be seen to be related to the disruptive activity. Before hunting, for instance, sharp-tasting fruit may not be eaten since a man may not expect to be able to exploit such fruit and sharp arrows at the same time. Similarly, both the size, shape and role in reproduction of eggs (be they soft-boiled or not) and testes are similar and both may not be capitalized upon at the same time. Not all activities have direct equivalents in the taboo system and so general self-restraint, particularly from sexual intercourse, is also frequently exercised.

Jane and Bali Kerei stayed behind to collect leaves, roots, flowers, fruit and fern leaves for lunch. We heard them approaching the *uma* some time before they came into view because Bali Kerei was shouting at the top of her voice about how expertly Jane was able to steer the dugout. They both disappeared into the inner room to join the other women and they must have found a lot to natter about because it took a couple of hours to make the sauce and boil some prawns.

In the afternoon, Jane and I were trying to take advantage of one of Toga's quiet periods by taking a nap when we heard a commotion from the river. We rushed (with care) down the narrow log from the hut and found Aman Uisak returning jubilant from an area of forest near his *sapo* with seven dead *bokkoi*. He had caught them in a *luluplup*, a sago-baited trap so effective at catching *bokkoi* and the occasional *simakobu* that its use is steeped in taboos and ceremonies. Up at the *uma* people were crowding round to see the haul and tales were told of how someone had once caught eleven *bokkoi* at once; or

164

was it twelve? or maybe thirteen? But no one scoffed at seven. Aman Uisak had already removed the heads so the skulls would hang in his *sapo*, drawn the carcasses and singed off most of their hair. Bai Uisak went through to the back room where most of the cooking was done, carrying the *bokkois*' brains in a palm-leaf bundle and the washed intestines almost overflowing a fair-sized cooking pot.

Aman Uisak walked down to the hut, followed by his children, and got himself ready to play the *tuddukat*. These are large horizontal slit-drums the largest of which, the 'mother', was about ten feet long and it was astride this that Aman Uisak sat. Next to it were two smaller drums, the 'children', about eight and six feet long respectively. Each drum was made from the trunk of a particular hardwood tree and in the centre of one side was a slit cut about one-third the length of the whole drum, through which the inside had been hollowed out. He took a long beater, a *tetete*, in his right hand and began to strike the drums one by one in a slow, deliberate tune.

The sound produced was surprisingly penetrating and the drums were used to pass tritonal messages to the surrounding *sapos* and even to neighbouring *umas*. Each drum is associated with a set of vowels: the 'mother' was sounded for i, u and ui, the 'middle child' for ou, o, e, ay, eu, and oi, and the 'small child' for a, ai, and au. Thus, if you wished to beat out 'it's a sunny day' you would hit 'mother', 'middle child', 'small child', 'mother' and 'middle child' to correspond to i, ay, a, i, ay. It could, of course, have been ui, ou, ai, u, o but this potential confusion is avoided by using a relatively formalized set of words and word order. When Reimar Schefold deciphered the musical code he checked it with the Sakuddei. They couldn't understand what on earth he was getting at for they have no script and no concept of vowel sounds. They don't rationalize the way they hit the slit-drums in the order they do and they can't sing the tune they are about to play.

Aman Bulit joined Jane and me and helped to interpret what we were hearing. The 'message' was divided into verses the first of which could be translated roughly (with some licence) as:

> This is a message for all who can hear,
> From the children of the Sakuddei *uma*,
> Those whose hearts so passionately seek
>   the wild animals in the hills,
> Those long-ridged hills, those high-peaked hills,
> You provide just what we desire, you my ancestors,

> You with the generous hearts who guard
> the animals in the hills.
> A song!

After the declaration of the *uma* name, the self praising and the thanks to the ancestors for their role in shooting animals, came seven verses all very similar to this:

> Come quickly, come slowly,
> View the victim of our hunt:
> A male *bokkoi* who calls '*wauk*' in the hills.
> That is just what we desired my ancestors.
> A song!

Thus each *bokkoi* was listed in turn and clan members not present were invited to join the forthcoming feast. At the same time it is hoped that those animals' souls who still live in the forest, and who can understand the messages, will be enticed to enter the *uma* for only then will their bodies present themselves to be shot.

The response to the call of the *tuddukat* was such that twenty merry men, women and children sat down together in the inner room that night to feed on the tender dark, lean *bokkoi* meat. Topoi Oggok and his wife ate only small pieces but with due regard to their senior years they had been reserved a generous share of the brains and liver which were soft and asked little of their sparse teeth. Later, after more storytelling and songs on the veranda, we went to our hut and discovered one of the disadvantages of telling everyone within miles that seven *bokkoi* had been caught. While we had all been feasting someone from downstream had rummaged through our packs and taken some of the tobacco and beads. As was typical of petty thieving on Siberut, only some of each commodity had been taken – no one would dream of stealing the last of anything. Losing things always made us sad, but in this case it was much worse since they were all gifts for the Sakuddei.

On our last morning I was tattooed. Up at the *uma* people gathered to discuss what tattoo lines they thought I should be given and I described our ideas to Aman Duman Kerei. From the middle room he collected a half coconut shell which was sooty on the inside from being hung above the hearth and into this he squeezed the sap from two plants that grew near the clearing. He coated the springy rib from a palm leaf with the black mixture and, as I stood before him on the entrance platform, he laid this as straight as he could from the base of my sternum to a couple of inches below my navel. The assembled crowd shouted out their approval and warned Aman Duman Kerei not to start tattooing on the sternum itself because it caused even Sakuddei men to faint. Having agreed on the basic line, it was decided that some embellishment around my navel was required. To print repeatable curves, Aman Duman Kerei bent the palm rib in two without breaking it and dipped the curved end in the black fluid. People shouted down each other's suggestions, putting forward their own ideas. Art by committee; it was easy to understand how Aman Duman Kerei had ended up with an area of tattooed doodles on either side of his knees. At length it was agreed that I should have four half circles around my navel with two dots in the centre of each.

I lay down with my back bowed over a large wooden platter and people craned their necks to see how I would react to the tattooing itself. Government officials, not themselves tattooed, had told me of the excruciating agonies of even a simple tattoo, and I was interested to see for myself just how it felt. Then it began; a sharpened, brass needle set in deer antler was coated with fluid and held just above the design and an *ariribuk* stick was beaten rapidly onto the needle holder as it moved slowly down the line puncturing the skin.

'*Malutlut aleh*? Does it hurt?' Aman Bulit called out.

'*Tak leu*, hardly,' I replied with some honesty and to the approval of the company. I was relieved, however, when after fifteen minutes the last puncture was made. I tried to sit up but I was held down and everyone started laughing. '*Tee, takpei*, it is not yet finished,' they shouted, because the needle had to run over the reddening skin still twice more. By the time the tattoo was finished and washed, the now blue weal was itching more than hurting. As was customary, I stood in front of the Sakuddei and they praised their handiwork and my nerve.

It was very sad to have to leave the clan, particularly for Aman Bulit. They had revealed so much to us of a caring family life, whose awareness extended to their whole environment. Aman Bulit and I gave them what remained of our stock of beads, all our clothes save what we stood up in, and most of our tobacco and food. Jane had already given Bali Kerei most of what she'd carried. The clan members sat in a subdued circle around the veranda and we said goodbye to each one in turn. We left the *uma* to the mournful, though strangely comforting, sound of soft moaning.

Aman Duman Kerei guided us over a ridge and down to a river to the west of, and parallel to, the Kuddei. We then struck north and crossed over into the Kaleat river basin. Aman Duman Kerei left us there and after he'd given Aman Bulit some rough directions on how to get further north and a quick lesson in the Paipajet dialect, we said more sad goodbyes.

The Kaleat had recently flooded and the tributaries that we had to cross to reach it were full of deep, sticky, soft mud through which we had to pull each other in turn. The fertile flood plains were all exploited, or clearly had been exploited in the recent past, for growing bananas and sago by inhabitants from Paipajet. This added further confirmation to our growing suspicions that all land on Siberut that is suitable for agriculture was already being used to some extent. Claims that logging helped to open up land for crops were fallacious to the extent that such land was largely unsuitable and any crops grown in those areas would almost inevitably fail after only a few years.

After several abortive attempts we crossed the Kaleat using me, as the tallest, as the depth sounder in front; if the water rose above my chest, then Jane and Aman Bulit knew they certainly couldn't make it. Aman Bulit found us a path through the forest that seemed to lead in the right direction and four hours later, just as we were considering building a shelter for the night, we came across ten little bamboo shelters covered with palm and banana leaves presumably built by Paipajet villagers visiting their fields by the Kaleat. We selected the two that looked the most sturdy, shaking them well to evict any lingering snakes and then stripped off to soak ourselves clean in a crystal-clear turquoise pool just downstream. Aman Bulit managed to light a fire almost immediately, using one of the more dilapidated shelters for fuel, and, before long, bamboos of our remaining soya and green beans were boiling furiously and sticks of stale sago were warming through. It was a spartan but convivial meal and we hoped it wouldn't be too long before we reached habitation.

The following day was plagued by no-through-gulleys, and flooded rivers. By late afternoon it was raining hard, and Toga was in an understandably filthy mood not helped by the fact that we had started rationing the few remaining bananas. We, too, were getting pretty hungry and despondent as we clambered out of a rising river onto a slippery flat rock by the bank. Suddenly, we all fixed our gaze on a cleft of rock to our left in which lay six *puruts* of sago. We didn't stop to consider who on earth owned them or what they were doing there; we just ripped off the leaves and thanked *Taikamanua* for his beneficence. I was chewing on the damp food, idly regarding the dripping scenery when I noticed something odd behind some fallen trees and branches some thirty yards downstream. If I hadn't known better I'd have thought it was the outline of a man wearing a loincloth holding a taut bow and an arrow pointing straight at us. Strange. It moved; it couldn't be; it was.

'*Aleh*,' I whispered softly but urgently to Aman Bulit, 'you'd better finish your mouthful quickly and use your best Paipajet language to tell the man behind that treefall not to shoot at us.'

'*Teelay*, where?' he mumbled through bits of half-chewed sago. 'Over there, just below where the bough rises out of the water.'

'*Teelay*,' he said. I was wishing he'd say something else when he waved his arm and spluttered a few words of reassurance and greeting. The man lowered his bow but as soon as Aman Bulit walked towards him, the bow was raised again. At length the man edged towards us and Aman Bulit kept up a one-sided conversation. '*Ba magilak bolaik*, do not be frightened; we are from Saibi Samukop. We are travellers looking for *beelow*. We have come from Sagulubbek after coming from Sarareiket and Silaoinan. We want to go home to Saibi Samukop but first we must go to Saibi Simatalu.'

The slender, middle-aged hunter accepted some tobacco, put his weapons away and told Aman Bulit how to reach shelter that night. We were just leaving him to strike up to the ridge top when the hunter called out to his wife. She emerged, trembling with either cold or fear, from behind a little bush above us. The poor lady must have been terrified and we wondered whether the man's failure to loosen an arrow in our direction had anything to do with the fact his wife had been so close.

The man's directions proved successful and we were soon made welcome in a small *sapo*. Sleep evaded me for much of the night, not because I wasn't tired physically, but because the sights and sounds of recent days set me thinking about Siberut's future. Acculturation was nearly complete for many of the island's inhabitants. Missionaries had gone beyond preaching that Jesus is the loving Son of God and attempted to ridicule and erode the traditional culture and substitute

German or Italian practices and values. We heard of one Italian pastor who taught that all beliefs in taboos were sinful and that if the men went out to hunt *beelow* no one would mind. Apart from missing the opportunity to use the taboo system as a means of explaining conduct in Lent, for instance, what he proposed was illegal and ignored the fact that a proportion of culture is separable from religion – after all, Harvest Festival is a Christian adaptation of a pagan custom.

But even if the missionaries are able to alter the Siberut religion and leave the culture more or less unscathed, the children would still learn at school that their traditional way of life is primitive. They then lose pride in themselves, their heritage disappears and many crafts, already dying, will be lost. In Saibi, for instance, there is only one set of *tuddukat* gongs left, made by Ohn's father, and they are old, split and seldom, if ever, used. Several months later Jane and I tried to enthuse Ohn and others to make a new set for the village for use on special occasions, but all claimed that they no longer had the know-how and, anyway, they couldn't see the point of making them. The great influx of tourists to Siberut, envisaged as the golden future by many officials, will never happen if all there is to see are disillusioned people sitting in villages. Western tourists are hungry to see simple alternative lifestyles that are close to Mother Earth, and cultures that have some element of mystery. Soon, these will not be found and the development policies are strangling one of the viable means by which Siberut could genuinely contribute to the Indonesian economy.

In fact, for those people interested in the Siberut culture, it is now probably a better use of time and money to visit European museums and to read the relevant papers in learned journals than it is to visit the island itself. A succession of professional and amateur anthropologists has removed most of the interesting artefacts from the houses. By itself this wouldn't have mattered had the culture been in a more robust state over the last few decades and the people still proud of their traditions, but the artefacts have either not been replaced or the substitutes have been made with much less care and skill. For instance, some of the southern clans which come in contact with the few tourists that succeed in seeing anything more than Muarasiberut, keep their tobacco and cigarette leaves in polished and carved coconut shells hung about their waists. These are an ideal size for a tourist curio and men have been persuaded to part with them. A glance around one of the villages now will show that the containers have not been replaced with similarly attractive coconut shells but with plastic bags stuffed into old sardine tins, tied around their waists with plastic string. People might be persuaded to continue their crafts if they weren't paid such derisory sums for the fruits of their labours.

Three days later Jane and I left Aman Bulit in Sirisurak and walked home wearily through the familiar, kindly forest in the south of our study area. It was getting dark and we had given our torch batteries to the Sakuddei, so we hurried along as fast as our reluctant, heavy legs would carry us. Suddenly Jane shouted at me from behind, 'Keep going, you've just trodden on a python.' The words and their order seemed so totally unlikely that I stopped and looked down the path behind me. Just three yards away was a large python curled up motionless on the ground. Jane made a wide detour around it and then stood next to me as we both admired the animal. It must have been fourteen feet long and the diffuse golden yellow hue along most of its length identified it as a large specimen of the so-called Short Python. We were surprised at how still it lay, but a flickering tongue betrayed the fact that it was, indeed, alive. It was the first time I had been able to contemplate a live python closely in its natural habitat. How aseptic and unimaginative zoo cages were. How distracting from the python's functional beauty.

Our camera's flash batteries were dead and I had intended to return with fresh ones to photograph the snake. However, as soon as we had pushed our way through our overgrown clearing, put Toga in her basket and flopped into the chairs on our veranda, the dammed-up feelings of sheer exhaustion and relief of being home swept over us and we slept and slept and slept.

Chapter Nine
Final Days

In my last six months of studying Sam and his family, I wanted to collect rather more detailed quantitative information than I had done previously. I needed to examine their forest more closely and tease out some of the delicate interactions I believed there were between them and their environment. So I was eager to get back to the *beelow* after a fortnight away buying stores on the mainland. The night had been cool and Sam didn't sing before dawn. I guessed that he might be sleeping near the bottom of the West Ridge and was rewarded by picking up the *beelow* soon after they had left their night trees. I tailed them cautiously for a few minutes planning to let them see me only when they were settled in their first fruit tree. From past experience I knew that after a break it always took them a little while to reassure themselves that I meant no harm.

I counted the gibbons as they swung into a fruiting fig tree; one, two, three – four? I must be mistaken surely? But four *beelow*, not three, had gone through. Who on earth was the extra one? It couldn't be Bess returned from some strange, long journey, surely? I walked to a position where they could see me plainly and their reaction changed my question from 'who was the extra one?' to 'who were they all?' They had scarcely looked at me before they all tore northwards, hooing with alarm as they went, a reaction Sam had never shown even after a month away from me. I rushed after the four disappearing black shapes, catching up with them just below the Glade, and they switched from giving soft hoos to swinging frantically around the canopy, sirening and giving alarm trills. At one point the female became so agitated and confused that she sang a great-call during which I was afforded an excellent view of her face.

Good grief! she was ugly. I may have been biased, but I'd always found Manai's and Bess's features rather soft and pleasing, but this poor lady looked as though her face had been knocked to one side, pulling and pushing her facial skin into unusual patterns. After about half an hour I somehow lost the group and so had a chance to sit down

172

and think. Where were Sam, Manai and Katy? Had these strangers ousted them in some dramatic territorial tussle? Had they been forced to give over some of their territory? Maybe Sam's home range had been artificially large after all. Why did this have to happen while I was away!

I failed to find the new group or any other *beelow* during the rest of the day but the next morning the new male was singing and I was able to sit close to his night tree and take down his song in shorthand. It was unique among all the *beelow* songs I ever heard by virtue of the fact that it was the most boring, uninspired, repetitive, uninteresting effort imaginable. Gone were Sam's exciting, flamboyant trills and subtle phrasal development. Instead, this male seemed to have no idea about expression, phrases or syntax, and I nearly cried at his incompetence and lack of style. Disappointment apart, the extremely characteristic, recognizable song allowed me to pinpoint where he had come from. When I had heard him previously I often thought to myself how lucky I was that Sam had such an entertaining song. This male had come from a territory that occupied part of the extreme north-west of the study area but why he and his family had come down here was still a mystery.

I followed him, the adult female, the small but independent juvenile and a subadult to the first fruit tree, an enormous spreading *tumu*, but I lost them soon after. It was just too depressing. I had spent over a year trying to get a *beelow* group accustomed to me and I felt that Bess's death had been quite sufficient a handicap to my study. There was now no chance of my completing the projects I had planned. I didn't feel I had sufficient energy left within me to cope with habituating another group and, in any case, I wouldn't be able to find the additional funds for the extra time in the field that would be necessary. For the first time in two years I had really felt like giving up and going home; I began to long for crisp autumn mornings, a solid and familiar civilization, chestnuts and crumpets toasted by an open fire, and the lights of Cambridge market on a late December afternoon.

It was back to the days of tramping the forest paths watching for movements, rushing after them and the *beelow* evading my efforts to keep up with them. But if you keep your eyes open in tropical rainforest there is always something interesting to see. Walking along the Main Trail after one particularly unproductive morning I noticed a peculiar loud noise like someone plucking a loose piano string coming from a minor gulley. Further investigation showed it to be coming from a fallen tree trunk, the upper end of which had become hollow and held a small pool of water. I peered down into the hollow and noticed a quick movement just as the otherwise incessant noise abruptly stopped. I inched my arm down the dark hole and at the first sensation of movement I made a grab and pulled out a copulating pair of *tuktuk'ake*, a type of narrow-mouth toad, barely one and a half inches long. They were a mottled browny-pink colour above with orange legs, and the male seemed far more intent on persuading the female to let him ride in the correct position than on escaping from me. My arm was covered in a white spittle which proved to be a soft, camouflaging cocoon in which the eggs were laid and where the young tadpoles would live. A week later the whole of the trunk's hollow was full of white foam and within a few more weeks all the miniature toadlets had disappeared.

While I was resting by the stream below the Fig Trail writing some notes, I caught sight of a *jirit* squirrel busily searching in the under-growth and coming slowly my way. I froze and watched it toss leaves to either side with its blunt nose raising its head every so often to give a high-pitched descending trill. It failed to notice me until each of its four feet were actually on my jungle boots. Then, either because of the unusual texture or because I may have moved slightly, it shrieked with surprise, jumped straight up in the air, and scampered off through the fallen leaves as fast as its short legs would carry it.

I wandered up the Fig Trail and on rounding a clump of *ariribuk* I saw, right in the middle of the path, a one-man shelter of tall palm-leaves below which were the remains of five empty *obbuks* of sago. Not more than forty yards to the east was the night tree Sam had used most frequently. To begin with I gave whoever had built it the benefit of the doubt; he might have been caught in a torrential rainstorm. But few storms last long enough for one man to eat five *obbuks* of sago and, anyway, Aman Mynoan's *sapo* was only five minutes brisk walk away. Judging from the few footprints we saw, local people rarely entered our study area, and when they did they tended to stay in the flat areas to collect rattan cane or to fish. I looked around for foot-prints. There were none up the slope but tracks led directly from the shelter to Aman Mynoan's *sapo*. Surely, this was it. Someone from the *sapo* had heard that we were away from camp, knew it would be safe to hunt, and potted poor old trusting Sam. It was unlikely that Aman Mynoan would have done the hunting himself but it seemed very likely that it had been his adolescent son. Later, Aman Bulit did some detective work for me and confirmed my fears that a single male *beelow* had been taken. I felt awful. I told myself that it was only because I had habituated Sam so completely that he had been vulnerable to hunting. Jane more or less consoled me by saying that any Siberut hunter worth his salt could shoot down any animal he chose but I still carried a nagging guilt around the forest with me. The fact that Manai and Katy appeared not to have remained behind supported my ideas on the nature of the territory. It had been Sam's. He defended it from other males both actively in occasional boundary chases and by singing songs as often as the weather allowed. Manai and Katy were simply tenants in the territory without even part-ownership. Manai's songs kept other potential mates away from Sam, and Katy's allegiances had always been first and foremost to her mother whether surrogate or natural. So, with Sam dead, the only option for Manai and Katy would have been to wander the forest until they chanced upon a single male holding a territory, as Sam had himself once been, and try to move in. No one could claim their journey through the gibbon territories with resident females would have been easy, but I hoped they were successful in the end. There was no point in trying to find them when I had no idea which way they had gone, or how far. No, I had to make the best of a bad job and get as much information as possible on the new *beelow*. As Jane put it, 'Think of it this way, how many field ecologists ever get the chance to observe a natural experiment? You have the same forest but different animals. Try to see if they behave like Sam, Manai and Katy.'

After a month of chasing the new group around their home range, they tolerated me until they saw me move. Finding them through the day was made considerably easier by the synchronous fruiting of two quite common trees, *tumu* and *alibagbag*. Luckily both of these grew in a relatively restricted habitat – on or near the break of slope. By concentrating my time in those areas I was able to watch the group for some time almost every day. This new group behaved in a very similar fashion to Sam and his group, and for some aspects it was possible to combine results. Between them, they ate fruit from fifty-eight species of trees and vines and although these could be classified into various groups such as 'soft and sweet' to 'rock-hard but with pulp-covered seed', they all contained some moderately tasty pulp. They rejected about one-sixth of the types of fruit available and these were either too large or extremely hard, containing nothing but bitter seeds, or too dry. It was possible, therefore, to predict reliably from examining fallen whole fruits whether or not the *beelow* would eat them when they passed through the tree.

By far and away the most interesting thing I found was that the area of forest used by Sam and his family was used in an almost identical way by the new group. Only once did Sikotkot, the male, venture beyond the boundaries patrolled by Sam and this one exception merely smoothed off a ragged corner. They travelled around the home range, not just using the same trees but using the identical branches to get from one to the next. Sikotkot slept in some of the same trees Sam had used for sleeping although there were apparently a large number from which to choose. The new female showed a similar lack of originality in her choice of sites for great-calling. All this suggested that there was a certain intrinsic significance to each part of the forest and it would be reasonable to view a gibbon home range as a home that has more or less well-defined rooms and corridors some of which have quite a specific use. The 'removal' of Sam showed further that the home can be vacated and reoccupied by strangers and it is obvious to a *beelow* how each room should be used, and where the corridors are. I wanted to pursue this line of argument further and so began a thorough survey of the trees in the home range, recording shape, height, dependent vines and epiphytes, and the local name given to me by Aman Bulit, Ohn and other friends. In this way I hoped to build up a picture of the physical and biological make-up of the *beelows'* forest.

When it came to analysing my tree measurements and identifications I found there were seven types of forest within Sam's and Sikotkot's home range. Some of these types, like tall dipterocarp forest and swamp forest were obviously distinct and Jane had found considerable differences in the abundance of squirrels and rats between them. But the five intermediate types were all rather similar to look at. Were the *beelow* sensitive to the subtle differences in structure and species composition? To my delight (and surprise) they were.

They also seemed to use different levels of the forest types for different purposes through the day and the idea of a house with a bedroom doubling as an occasional sitting room and with a cellar that is visited only rarely and for a specific purpose was not wholly inappropriate. This leads one to suspect that although the position of home ranges and territories is maintained in part by the activities of the surrounding groups, each area in an uncrowded population provides

176

all the elements that a *beelow* group requires, and the patches of forest between home ranges which never seem to be used at all by *beelow* probably provided a resource that was already abundant within the groups' 'homes'. There is, therefore, an unexpected permanence about the way *beelow* space themselves and, if the study area is not logged, I'm sure that I could return in twenty years' time and find more or less the same boundaries I found during my project, although the groups defending them would be different.

The final phase of our surveys for the World Wildlife Fund was to investigate the effects of commercial logging on Siberut. Siberut's forests, in common with other tropical rainforests, are under tremendous pressure from human disturbance. It has been calculated that the area of tropical rainforest lost annually through commercial logging and slash-and-burn cultivation is some 43,000 square miles. Such a mind-boggling number is hard to comprehend and it is helpful to think of it as the area of four *beelow* home ranges or London's Hyde Park being lost every seven minutes.

Loss of tropical rainforest also occurred in prehistoric times. New evidence from fossil pollen grains suggests that during the last few million years the composition and therefore structure of rainforests have been in a state of considerable instability. When the temperate regions experienced ice-ages, the tropics were subjected to periods of comparative dryness. As the area of humid forest retracted, so the fringes became deciduous woodland and other types of warm- and dry-land vegetation. But the present rate of loss is magnitudes greater than it ever was in the past and there is probably less tropical rainforest standing now than there was at the peak of the past's most adverse dry period. Between the ice-ages tropical rainforest arose again, but these fluctuations took hundreds of years and each began from some form of natural vegetation rather than from secondary growth on eroded slopes.

When we began the surveys, barely one and a half per cent of Siberut's area had been declared a Nature Reserve and over ninety per cent had been conceded to four logging companies. These were owned by foreign consortia from the Philippines, Singapore, Malaysia and even some Russian interest, but because of the rise in timber taxes, these were being sold out to Indonesian businessmen, well-to-do government officials or local government.

We concentrated our studies on logging around a fairly typical, medium-sized camp above Totoiet Bay on the east coast where the muddy shanty town of long, plank-walled dormitories were full of the sound of blaring cassette players. We were made very welcome by all and in particular by Chai Gan the Chinese Malaysian camp manager who was more than grateful for any distraction from his tedious job. The fifty or so men at the camp lived quite well and fresh food was brought from the mainland every three or four days on the company boat. It was cooked in magnificent Chinese home-style by two West Sumatran ladies and Chai Gan reminded us incessantly to make sure we 'took some lunch' before we went out to work and as soon as we returned. After 'lunch' on the first night Jane remarked to him that the work force seemed to be comprised entirely of Malaysians, Bataks from North Sumatra and Javanese.

'It's like this-*lah*,' he told us, after calling for another pot of coffee, 'we like the village boys but I didn't hire many boys yet, isn't it. We teached some to drive tractors and the boys learn well-*lah*, but usually we use the Siberut boys only for surveys in the forest, isn't it. They find us big trees and we pay them; less than the regular boys but no problem-lah. But,' he continued rather more candidly, 'they didn't become very good workers yet for cutting timber, isn't it. They come, we give them work and after a month-*lah* we pay them Rp20,000 (about £20), then they rush back to their village until they have no more money left and they come back; many problems for us, isn't it.' He laughed and poured some more coffee. With the absence of any organized market economy on Siberut, any time a man takes away from his gardens means that chores mount up and he has considerable work to catch up on. We also heard that local men were rarely employed because the act of felling a tree for no readily apparent reason other than money is anathema and would normally have to be preceded by ceremonies and divinations. Such practices hardly lend themselves to a profit-seeking operation.

We found that attitudes to the loggers varied tremendously. Women who brought chickens and fresh fruit and vegetables to the camp cooks, from a nearby village away from the logging roads, welcomed the money and goods they received in return and had few grumbles. Youths from Sirisurak, as yet quite unaffected by logging, helped on a logging survey and felt that, as the companies gave so much in the way of money and perquisites for so little work, logging had to be a good thing. Villagers in the Sirilogui region had sought the

178

help of chainsaw gangs to clear the surrounding hillsides before the camp was abandoned. When we visited the area eighteen months later, erosion gullies ran down some of the slopes like obscene lacerations but sugar cane, cassava, bananas and pineapples seemed to be growing everywhere else. The villagers were in two minds about the benefits; they were able to grow more food, that was true, but there hadn't exactly been a shortage beforehand and they couldn't sell the surplus because there was no market. They were finding that fishing in the silted streams was less successful and that the areas which had been cleared and cultivated first were already becoming less and less productive. Also, finding suitable trees for dugouts now necessitated a long trip to the headwaters or to an area in the north. In the south of Siberut where the history of logging was longer, local ire and hatred had been raised because of the way traditional ownership rights to land and trees were ignored, compensation was miserly or nonexistent, and venereal disease had been introduced.

Life in the logging camps palls rapidly for the men, with the heavy machinery out of action two days in three. When wet, the unresistant shale and marl that dominates Siberut soils is totally inappropriate not to mention dangerous for the large tractors used to transport personnel and timber. Thus work is halted when rain falls and work on the following day is generally abandoned as well because the roads remain slippery and muddy. This is one of the reasons none of the companies had been able to make a profit from logging on Siberut. Another is that transporting the timber is so difficult. Most of it is destined for Japan, Korea or Singapore and the position of Siberut relative to Sumatra involves the ships and large tugs towing the cumbersome pontoons laden with logs in long trips round either the northern or southern tip of the huge island.

We conducted two types of surveys at Totoiet; one along logging roads watching out for any signs of primates or deer and the other hacking measured transects through the tangled, mangled forest to calculate the number of trees felled per acre and to count the density of *jirit* feeding trees that Jane was now able to recognize with ease. Primates of all four species and squirrels still inhabited the logged forest but we only ever saw singletons of *beelow*, *joja* and *simakobu*. No one could deny that logged forest is able to support some primates, but we still don't know how they are affected by disturbance in the long-term; their group formation is clearly upset and reproduction in the intact groups may also be reduced or disrupted completely. Nor do

179

we yet know how they cope with a forest that has been chopped into islands by wide roads. All species could cross the roads if they were leaving a parental territory, for instance, but I cannot imagine a *beelow* group incorporating a possibly hazardous walk across a road in their daily routes around their own home range. For arboreal primates a disturbed canopy reduces the number of alternative pathways through the trees and they suffer more from human hunting in such areas where they cannot flee so quickly. We found a number of arrowheads along our transects that attested to this. There is also no clear indication of how the loss of a primate species in an area would affect the trees. It is known, however, that some plant seeds have to pass through an animal gut before they will germinate and it is quite conceivable that local extinction of a primate would lead to a similar extinction of plants.

It is an unfortunate fact of life that the smaller and more isolated an island is, be it an oceanic island, an 'island' of forest in an ocean of agricultural land, or a patch of forest surrounded by logging roads, the fewer species it can support. At the time of writing, about sixty-five per cent of Siberut is suitable for those species (the vast majority) dependent on more or less undisturbed forest. Their island is thus a third smaller than it was before logging started ten years ago and so the extinction of some species is virtually inevitable in due course. But which species? It is not possible to say for sure and no mass extinction will occur overnight, but it is most likely to befall the less adaptable species such as the monogamous *beelow, joja* and *simakobu*.

It didn't take many days before we discovered that this logging company at least was paying little regard to their concession agreements. Trees below the twenty-inch minimum diameter at breast height were being felled, roads ran beyond the concession boundaries, trees were felled near rivers and no tree appeared to be on too steep a slope for the chainsaw gang. Logging itself need not be disastrous to a tropical rainforest if it is conducted with due regard to simple principles. In this way large trees can be harvested on a regular, possibly forty-year, cycle. This is what the concession agreements are supposed to provide but the forestry officials and policemen sent to check on their implementation are paid very little and probably see no harm in allowing their generous hosts to take what trees they like.

Our transects revealed that on average four trees were felled purposely in each acre, about ten per cent of the available timber. That doesn't sound so very many but the enmeshed vines and roots cause many other trees to fall. In addition tracks must be cut to reach and extract the timber, resulting in over half of what remains being destroyed. Indeed, these figures exclude those trees felled to make way for the major roads and the extensive marshalling yards and those not pulled or pushed over but just left maimed, wounded and laid open to fungal attack. There are also trees, shrubs and herbs adapted to shelter and low-light levels which, deprived of these conditions, are soon killed by storms or smothered by vigorous, light-loving secondary growth.

One of the most disturbing facts we uncovered was that roughly a quarter of all trees felled are left where they topple. The chainsaw gang must work out how each tree should be cut in order that it will fall where they wish. On one occasion we watched terrified as a tree weighing about fifty tons leant the 'wrong' way, causing the massive trunk to crack open wildly from the base as the split raced up to the crown, and then the tree dropped to one side. The gang were lucky to escape with their lives; sometimes they didn't. Other trees were felled perfectly but were found to have rotten unsaleable hearts. So they were left; a wasteful mistake.

The concern over loss of forest habitat is not purely esoteric. There are pressing human social and economic reasons why the present prevailing policies of felling and selling to raise large sums of foreign exchange and investment rapidly are ill-considered. It cannot be denied that the income currently received from logging operations is easily thought of as a panacea for a less developed country's seemingly endless catalogues of money crises. But most of the advantages are only short-term (though long enough to make a few individual bank accounts look very healthy). If a forest is felled selectively for the few species sought commercially, and then slash-and-burn cultivators clear what remains and scratch a subsistence living from the thinning, tiring soil until it is exhausted — what then is the value?

In contrast, fewer logging companies could extract less timber on a forty-five year cycle and use parts normally discarded such as boughs and bark in those industries using lignin and cellulose, the forests' most common substances; indigenous inhabitants could exploit the forest products at a controlled, sustainable level; water and climate regimes could be maintained; fertility of flood plains could be ensured; the majority of species would remain, tributes to Man's slow-evolving wisdom and restraint, and could be exploited for foreseen and as yet unimagined needs of Mankind. Siberut has, at a conservative estimate,

181

about 1500 species of flowering plants, more than the whole of Great Britain which is fifty times the size, and of the fraction already collected there some have important potential uses as gene stocks for fruit and root crops, spices, seed grains, animal forage, medicines, fibres, gums, dyes, resins and insecticides. Many others doubtless exist; a *sikerei* and his herbal cures could keep a botanist, an analytical chemist and a medical scientist busy for years. Many of his concoctions may have no measurable effect at all on man – but some may. And who would wish to pass over a drug that could cure an as-yet-incurable disease or condition when hospitals already use drugs that originated in the forests of South America, Africa and Asia? Some of these are used in a refined state, whereas others have been the clue to manufacturing a derivative that avoids side-effects inherent in the wholly natural drug.

But one doesn't need an economics degree to predict that such an Utopian organization of resources would result in greatly increased timber prices. Exactly! this is the constantly recurring problem of world development. Why should timber be artificially cheap at the expense of the developing country producing it? Why should easy investment be allowed in the gross disturbance or even destruction of the world's richest and potentially most useful ecosystem and in the inevitable extinction of hundreds of some species as yet undescribed and unevaluated? If the real value of a tree that once helped maintain the world's climatic *status quo* by absorbing the sun's rays and increasing quantities of carbon dioxide, that protected a fragile soil, moderated the flow of water to the river, and provided living space for epiphytes, support for rattans and food for animals, is greater than the present market value, then that is the price that should be paid.

A few weeks after returning from the logging camp there was great excitement when one of Jane's rat traps on the East Ridge caught a hitherto undescribed species of treemouse. It was just under eight inches long, five inches of which was tail, and had whiskers as long as its body. It had the big, black eyes of a nocturnal animal and was mid-grey above, white below, with the fur at the end of the tail covered in the softest, whitest hair imaginable. Its relatives in Malaya are found frequently in nests inside lengths of bamboo but since there was no bamboo on the East Ridge, it probably nested in hollow tree trunks. Not only was the treemouse new to science but it was also new to all the people who visited our camp and so had no vernacular name. Aman Bulit, however, considered the situation sagely and decided it should be called *sipupaipaijutoh* or 'the one with a tail like hanging clubmoss'.

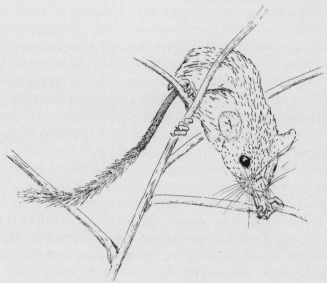

Sipupaipaijutoh

A few days after the treemouse was found, I stayed in camp to catch up on paperwork and Toga seemed to be off-colour. When Jane returned from a day of checking her traps, I said, 'Come and see if you can tell what's up with Toga. She's really hot and has been clinging to me like a leech since you left.'

Toga had eaten nothing all day and been reluctant to eat even though her urine showed clearly that she was becoming dehydrated. Jane took Toga's sweaty hand in hers and was greeted only by a sorrowful glance from a pair of brown watery eyes. Toga had been her normal ebullient self the day before but now all her natural zest seemed to have drained out of her. We decided to start her on a course of antibiotics, estimating what we hoped would be a reasonable dose, but had considerable difficulty persuading her to take anything. Early the next morning I found her lying dead in her basket. I cradled her body in my arm, took it into the forest and laid it on the ground to take its place in the natural cycles of the forest whence she had come.

Toga's sudden death after seven months of caring for her day in

183

and day out saddened us greatly and left a painful gap in our lives. Raising her had been a wonderful experience for both of us. We also felt a measure of relief, however, because with only six weeks to go until we had to leave Siberut it had become clear that Toga would still be too young and clueless to live in the forest on her own or in a group.

During our last weeks at Paitan we worked ourselves silly packing, clearing rubbish, sorting out which of our local guides would get what, and doing all those jobs in the forest we had inevitably left until the last moment. Out of these was looking at rat activity between dusk and dawn. I'd built a small shelter of bamboo covered with banana leaves at a bend in the Tolailai stream by Jane's four-acre trapping grid, and we stayed there on two sets of two consecutive nights, checking each of the 120 traps every two hours and releasing any occupants. Whereas dawn was always an exciting time with so many animals waking and the prospect of a sunny day ahead, dusk was cold, damp and mysterious and we looked forward to total darkness. Only then did the dense coatings of fungal strands on dead wood and across the leaf-mould luminesce as they digested the rotting material, and fireflies flashed and danced in the treetops. We were blessed with largely dry weather and we kept a good fire burning on the light-brown pebbles of the sloping stream bank. This had the combined functions of a welcoming and guiding light after our half-hour trap inspections, a source of smoke to ward off biting sandflies, and an aromatic cooker. On the second occasion, Bai Taklabangan gave us an *obbuk* crammed full of *subbeh* – mashed taro balls rolled in grated coconut – and while these were heating up we set our two fish traps in pools upstream using the pith of a sago palm after *tamara* grubs had eaten their way through it as bait. On the way back to the rude shelter we caught four slippery, seven-inch *teilek n'oinan* frogs that looked deceptively like the large pebbles on which they sat, and later we had a delicious midnight feast of *subbeh* with roasted frogs' legs and crayfish from the traps. We stored the few fish we'd caught and rebaited the fish traps in the hope of a roast-fish breakfast.

The red forest rats Jane was studying closely were most active in the early evening and early morning with the three hours either side of midnight producing few occupied traps. There were a few trap-happy individuals, however, that seemed to have nothing better to do all night than get caught. We suspected they hid behind a nearby tree after being released, waited until we had rebaited the trap and then returned for another meal of the sago and peanut bait, thereby releasing the trap mechanism again. Over the past month some of the young rats had been caught every night and seemed to be losing condition on the rather unnatural diet; Jane had therefore had to close all the traps for a few days so that the rats had a chance to eat forest food and maintain their social positions.

On our last day at Paitan, Jane and I went into 'our' forest for the final time – without notebooks – to enjoy it for what it was rather than to study it. We had grown to know it extremely well and although it had frustrated us, hurt us, deceived us and outwitted us, we had also been marvellously happy there. We were graced with a magnificent morning; the sun rose from a wet, hazy horizon, dissolved the mist and then struck the proud dipterocarp trees, setting them aflame with golden light. At seven o'clock we sat on Summit Three and a female *beelow* began to great-call from the East Ridge. Before long she had been joined by others and the living galleries and naves were filled with thrilling, ethereal-sounding music.

Loga

We walked down the Main Trail together for the last time but at Summit One we halted abruptly when we noticed a *loga* near the ground not four feet away. It bounded off up to the treetops where we saw another, then another, and another of these usually solitary squirrels. In all we counted eleven of them, following each other round a complicated circuit that brought them to where we'd seen the first *loga*, up saplings and vines, across a gap, along a broad mossy bough of a large gnarled tree, and down the vine- and creeper-covered trunk of its neighbour. Round and round and round they went, loping bushy tail after loping bushy tail, chucking and whirring, for close on two hours, giving us our best-ever view of these squirrels.

What we had been watching was known locally as a *sejeru loga* and Aman Bulit suggested that it was a ritual in which perhaps a dozen male *loga* come together to chase and attract a single female. At the end of the display the female 'chooses' a male and the new pair go off and start building a nest together. Our observations were never clear enough to sex or identify individuals but it certainly seemed plausible. Between us, Jane and I had watched three other *sejeru loga* and we were satisfied that they weren't just chance meetings of a large number of animals but it would be difficult to decide categorically what function they served.

On our final day on Siberut, I visited Totoiet again while Jane stayed at Muarasaibi with Ani, the Sirisurak schoolteacher's wife, to look after the few belongings we were taking off the island. Ani had always been very friendly, particularly towards Jane, and she had helped us and entertained us on numerous occasions when we were stuck at the coast waiting for a tricky wind to blow itself out. Her most memorable remark, made when she and Jane were sharing the floor in her shop, was that the reason Siberut children wore such broad, sunny smiles all day was because their parents always made love in the open air. News had reached me that the logging company at Totoiet was blazing a new road through the forest even further inland and I wanted to watch, no matter how sad it would make me. Mr Gan took me eight miles inland and left me to walk towards the sound of a throbbing, straining diesel engine alone. The sight was ridiculously incongruous: there, on a forested ridge-back barely distinguishable from any I had seen during the hundred miles of forested survey walks, was a bright yellow tractor with a rain of twigs, branches and leaves falling on its cab. It was adorned with streamers of several vines that seemed to be pulling it back but it roared forwards regardless, pushing undergrowth relentlessly beneath its cold metal tracks. When it came up against a tree of a size I would have regarded as immovable, the engines merely whined more frenziedly and the irrevocable and undignified fate of the tree was settled. What wrenched my heart more than anything else was the cock-snooting, mocking way the machine so effortlessly toppled the painfully inadequate vegetation, the way evolution had equipped the trees with any number of defences against pests and diseases but nothing that could fight back against the bulldozer.

I doubt if anyone on Siberut will ever throw their arms around trees in front of the loggers' saws and bulldozers as do the 'tree-huggers' of northern India whose subsistence economy also depends on their forest. Nor do I think they would ever start planning midnight raids on logging camps to pour sugar into the fuel tanks. Their traditional attitude to life is largely fatalistic for they have rarely had the means of altering the course of events that affect them. When Jane caught typhoid, for instance, it was unbelievably hard to get even good friends to help paddle her downstream on her way to a mainland hospital. Their response to desperate situations has always been to ask help from the spirits and then wait to see what happens.

The bulldozer ploughed on and I could do no more than stand in dumb grief. Another tree, a *popokpok*, twenty-five yards from me·was in the way; it shuddered when it was struck the first time, and then it began to lean. Suddenly a black animal leapt out of a hole in the tree and glided towards me. It swooped to within six feet of my head, blotting out the sun for an instant, and landed a few yards behind me. It was an enormous, jet-black flying squirrel, a good yard from nose to tip of tail with a body the size of a hefty cat. The 'wings' folded up into place and wobbled at the squirrel's sides as it ran, terrified, into the despoiled undergrowth. I knew immediately what it had to be. People had been telling us about a giant flying squirrel called *kaloloklok* ever since Jane arrived on Siberut and declared an interest in squirrels. On surveys guides had sat bolt upright during the early evening and told us to listen to its bubbling call – '*kalo – lokloklokloklok*'. We had once even heard the call behind our house, but we had no real evidence that it wasn't a night bird or even a frog. And there it had been gliding over me. Yet another undescribed mammal, a dramatic symbol of the untapped riches of Siberut's diminishing forest. I felt like yelling to the bulldozer driver to stop — but he wouldn't have heard me; worst of all, he wouldn't have cared.

Chapter Ten
Epilogue

Siberut's greatest asset in the search for a secure and realistic future must be that it is a relatively small island. It is, therefore, a tangible, comprehensible unit, largely independent of the mainland, which can be managed as a whole for the benefit of wildlife and people alike. This enviable situation thus thrusts Siberut forward as an indicator of the commitment of Indonesian governmental and non-governmental organizations to a future based on sound conservation principles. Siberut has so much going for it – its size, its high percentage of remaining forest cover, its low density of human population, its tourist potential, its intense scientific interest and the attention of, and potential aid from, a plethora of international organizations. How could it fail!

But first things first; a plan was required. One of the 'conservation' plans that has done the rounds of the local government offices in Padang over the last decade is to make the south-eastern peninsula of Siberut into a nature reserve. The rest of the island could then be cleared of trees and all the *beelow* and monkeys caught and taken to the reserve. Indeed, displaced primates from the southern Mentawai Islands could also be taken there and channelled through a rehabilitation centre where tourists would pay to see them. The tourists would have little difficulty reaching this reserve because it is situated near Muarasiberut and, *voilà*, everybody's happy.

The World Wildlife Fund was invited by the central government to produce an alternative plan. We were asked to do the groundwork, but then the results of our surveys, of our academic and other studies, and of the published work of other scientists and government officials were pulled together with the invaluable assistance, and under the overall direction, of Jeff McNeely, the World Wildlife Fund's assistant representative in Indonesia.

One of the central themes of the eventual report had to be to show that the goals of nature conservation and sustainable human development need not conflict. But before planning the development of the people on Siberut it is useful to examine one's own motives very closely. The desire to preserve the people as living museum pieces, representing much of what we value and envy in terms of lifestyle, is very strong. I certainly find fully-tattooed Siberut women with chipped teeth, a row or two of colourful glass beads and a wrap-round skirt, extremely handsome. Likewise I find the men with long hair and bark loincloths extremely impressive and I felt tremendously affronted once when I watched one such couple have their hair cut and their bodies covered with ill-fitting, somewhat decrepit Western clothes. But they had chosen to do this because they wished to move from Simatalu to the more westernized Saibi area. Similarly, Ohn had once taken me on one side to explain that his wife would soon be having their third child who they wanted to call by a European name, and so he asked me what I would suggest. I couldn't see what on earth was wrong with such lovely sounding names as Tengatiti, Taklabangan, Tupilaggai, Agistinu, Titikmanai and Takenukta. But he was adamant; at any one time there could be only one person with any name in the river basin and they had run out of suitable names. I was absolutely certain he could have found some local names that didn't have owners if he had tried, but I gave him a long list of boys' and girls' names all the same. He had made the choice to find a European name and what right had I to alter his own independent decision? The people have a right to choose the pace and direction of their own development even though fashionable trends may seem to go against our own ideals.

Central to all the elements of coordinated development on Siberut had to be a system of land-use zones. The first priority was to define the boundaries of a nature reserve to include as many types of undisturbed vegetation as possible. From this it was clear the reserve would have to be primarily on the western side since the east had suffered the greatest logging and supported most of the human population. Now, if logging were allowed to continue unchecked, it is more than likely that little of Siberut's forest would still remain intact. But rather than wait to be left with pocket handkerchief areas scattered all over the island, we suggested a single, continuous Nature Reserve Zone occupying an eighth of the island, within which all commercial and non-commercial exploitative activities would be prohibited. The chosen region contained excellent forest, but was also suitable because it was comparatively distant from any established habitation thereby minimizing any hardship or inconvenience caused by its restrictions.

We then devised a second area called the Traditional Use Zone. This occupied about one-third of the island and lay mainly around the

Nature Reserve to act as a buffer and as a reservoir of the forest resources required by people on a non-commercial basis. Commercial use of any type would be prohibited. No new buildings would be erected in this Zone and only small, traditional (unburnt) clearings would be permitted although existing sites could continue to be used. Hunting of *bokkoi* and *joja* with the traditional weapons would be allowed since, of the four primate species, they are the best able to cope with human hunting pressure. Even so, their numbers would have to be monitored in case it is necessary to give them total protection, and all kills would be subject to clan quotas; all skulls hung in houses would have to be registered with rangers from the Conservation Department. Continued total protection for the *beelow* codifies the traditional taboos and is essential because, in the absence of any protection, they are likely to suffer severely due to the ease of stalking them during their long, loud songs.

The remaining half of the island would include the regions that support the majority of the population and would be called the Development Zone. Some of it has already been logged and it provides the best potential for future logging and agriculture. Moderate, carefully controlled, selective logging would be allowed, for this need not be totally incompatible with the overall conservation objectives, and small plantations of spice trees could be encouraged to provide cash crops. With all these developments, however, restrictions would be imposed on the maximum slope of hillside that could be cleared in order to limit erosion which not only impoverishes the soil but also silts up the rivers and smothers coral – both of which result in poorer fish stocks. All new houses, gardens and villages would be confined to this zone and hunting regulations would be the same as those in the Traditional Use Zone.

Even before the Masterplan was published, the boundaries of the proposed nature reserve were declared officially by the minister of Agriculture, and this decision gave a great boost of encouragement to those involved with Siberut. Also Dr Emil Salim, Indonesia's Minister of State for Development Supervision and Environment, agreed to write the foreword to the Masterplan. In this he states that the Masterplan is:

> an important new step in Indonesia's efforts to link conservation of the environment with socio-economic development. . . . [It shows] how nature conservation can take its

appropriate place in integrated land-use planning . . .[and] how the values of the Siberut culture can be maintained while it is gracefully brought into the mainstream of Indonesian society and the natural environment is saved as well. . . [The] use of conservation techniques for sound socio-economic development is a new approach which, once it has been shown to be viable on Siberut, can have many applications elsewhere.

The Masterplan, with English and Indonesian texts side by side, was published in a blaze of publicity in March 1980 and front-page headlines such as SIBERUT MUST BE FREED FROM LOGGING CONCESSIONS appeared in Jakarta newspapers. The English language *Indonesian Times* even serialized the entire plan. At about the same time, Art Mitchell, the man selected by the World Wildlife Fund to help implement the plan, arrived in Indonesia from a conservation education project in East Malaysia. His passage through the tortuous corridors of bureaucracy in Jakarta was greatly eased by being able to present copies of the Masterplan to every government official he met, and its aims won wider and greater acceptance. Everywhere he goes on the island and the mainland he distributes copies of the Masterplan, pointing out the clear support given to the ideas by Dr Salim, and he has just begun to give more formal presentations with slides, posters and other visual aids to whole Siberut villages. In these he explains the regulations of the reserve and the reasons why they are necessary for, if the reserve is to be truly effective, the people will have to support the regulations. It is clearly important that they are shown that the rules will not be particularly restrictive to their traditions of wise land use, and equally important to show that some change is inevitable. The forest takes such a major role in some people's lives that many find it hard to believe that things will ever be very different, although when pressed will admit that some animals are more difficult to find now, and in some areas dugout trees are difficult to find. Art is also compiling a book of traditional animal stories illustrated with children's drawings. This will be in Indonesian rather than their own language so that it can be used as a language textbook in schools.

In the middle of 1980, exactly two years after Jane and I left Siberut, I was offered the chance to return there for a week. Art had kept us in close touch with developments but I could hardly turn down the opportunity of meeting old friends again and of seeing the situation for myself although I knew I would be saddened both by the brevity of the trip and by the changes.

When I arrived in Padang my optimism about Siberut's future was rudely jolted. Art told me he had recently discovered that within the Nature Reserve Zone 5,000 acres had been logged, 750 acres were in the process of being logged and a further 5,000 acres were due to be logged that year; and that within the Traditional Use Zone, 6,000 acres had been logged – a total area fifty times the size of Hyde Park. How could this have happened if the Agriculture Minister had declared the reserve official? There are two reasons: first, there was an interval of nearly two years between our defining the reserve boundaries and the date of the declaration – though this was no fault of the Minister's; second, despite the fact that the boundaries were official, it was still necessary for letters to be signed by the Director-General of Forests to cancel the logging concessions inside the reserved zones and this was not being done. Art and local officials of the Conservation Department were therefore powerless to influence the decisions of logging companies on Siberut. This caused friction between the World Wildlife Fund and the Conservation Department since their co-operation depended on the latter's ability to maintain the integrity of reserves. If this was not possible then there was little point in the World Wildlife Fund pouring charity money into supporting a lame duck.

Art saw the immediate priority for conservation on Siberut as moving the boundaries of the reserve to exclude the logged forest and include other areas of undisturbed forest, so maintaining the reserve's original extent. All this will require time, patience, and high-level representations but I have little doubt that there will, in due course, be a viable reserve on Siberut that is safe from the loggers' saws. Indeed one of the most hopeful signs I saw on my brief visit to Siberut with Art was along part of the ridge above the Paitan River that Jane and I had used when walking between Sirisurak and our camp. A wide reserve boundary, marked every half mile by a large mound of earth and a stone, and every hundred yards by a red painted metal plate nailed to a tree, is being cut by the Planning Unit of the Forestry Department along the route we had suggested. They are concentrating on the eastern-most boundary because once there is an obvious line it becomes far more difficult for a logging tractor to enter the reserve with any tenable excuse. In some areas it will be touch and go whether the boundary or tractors will arrive first, but given Art's determination and enthusiasm and Dr Salim's continued support I maintain a general optimism for the *eventual* integrity of the reserve.

I wish I could feel similarly optimistic for the people. Art told me that the resettlement of inland villages towards or on the east coast was continuing apace. The resettlement programme had begun slowly ten years before with the aims of bringing people out from the 'rigours and hardships' of the interior and of facilitating access and control by the authorities. Katinambut, the village of nearly twenty *umas* where Jane and I had once stayed and watched *sikereis* dancing and singing, no longer existed; the *umas* had been left empty and rotting. Worse still the *sikereis* had been stripped of their finery.

Just before I arrived back on Siberut, the mayor of the southern half of the island had produced an edict requiring all 319 *sikereis* within his jurisdiction to hand over all their ceremonial paraphernalia such as bells, beads and head-dresses, in an attempt to stop ceremonies being held. The mayor recognized the role of *sikereis* in healing and pronounced that he was happy for this to continue just as long as they only dispensed herbs. Even if one can't understand how important the accoutrements are, it is obvious that patient psychology plays an important role in healing, and the bells, songs and dances aid the process. The confiscations also deny people the right to keep alive their traditions. In addition, the mayor decided that all monkey skulls kept in some of the southern-most villages should be burned, and he generally seemed to be attempting a single-handed destruction of the Siberut culture. The desire to 'develop' the Siberut people is very strong among local government offices, however, and this was reflected in the way the main Padang newspaper hailed the mayor's actions as a grand step in bringing the people to civilization. All this was painfully reminiscent of the authorities' harsh attitudes in the late sixties and early seventies. Then, men were forced to have their hair cut, their headbands and beads were confiscated by police and, in some areas, their pigs were shot by intolerant Muslim soldiers. Those inhumane and unnecessary acculturation practices eventually ceased under the orders of the previous provincial governor.

At length, news of the mayor's edicts reached the ears of Dr Salim and he lost no time in sending a terse telegram to the governor of West Sumatra telling him that what the mayor was doing was illegal, should stop immediately, and that the person responsible for the initial orders to the mayor should be reprimanded. Nothing had happened by the time I reached Siberut (save meetings to discover who had leaked the news to Jakarta) but the *impasse* was expected to break at any time.

I managed to see Aman Bulit for only two hours because I had a restricted schedule and he had been away attending a wedding feast for his sixty-year-old uncle. Aman Bulit still had a gleam in his eyes, the warmest of hearts and the widest of grins, but he lacked some of his former sparkle. He appeared heavy laden and as we spoke about the problems of Siberut he became increasingly morose and confused about the future. In the two years since we had met, he (and all the other friends I spoke to) had become newly aware of all the threats to their settled, predictable lives. Logging tractors were fast approaching their lands and there were rumours that the inhabitants of Sirisurak would soon be forced to move to a coastal resettlement village. Aman Bulit had heard about the few other such villages in the south, and knew that there was currently a 7pm to 7am curfew in operation making it impossible for people ever to get up to the *sapos* or *umas* in the headwaters to tend their pigs, cut and make sago or collect rattan. He looked at me with a wistful smile and said, 'We shall need to buy an aeroplane if we are to beat the curfew.'

Aman Bulit and I held tight in one long last embrace on the beach at Muarasaibi as the motorized dugout came to take me south. Then I sat in the bow holding on to steady myself in the irregular swell as the engine thrust us forward in a corona of spray. As he and I continued to wave to one another, heavy grey clouds gathered over the Saibi and the scene seemed peculiarly symbolic. I could leave and worry – he had to stay and try to adapt.

I stayed for a few days in Maileppet, a village about a mile to the north of Muarasiberut, which is the centre of Survival International's project for socio-economic development on Siberut. Considering the general approval that met the mayor's actions, it is hardly surprising that a project which aims to maintain the local culture and traditions at the same time as bringing the people into the mainstream of Indonesian society, met with a great many problems. Thus, it took Gerard and Jes Persoon, the handsome Dutch couple whom Survival International chose to execute their project, seven months of repeatedly visiting a succession of offices in Jakarta and Padang before they were eventually allowed to start work. They finally arrived on Siberut just a few months before Art Mitchell.

One of the project's declared major aims is to introduce the people to alternative protein sources and to improve existing ones. Buffaloes, goats and pond fish have been suggested as new sources of protein but Gerard now feels that only the last of these has any real future. Buffalo are expensive, would be difficult and costly to transport, may seriously degrade the river banks and they provide too much meat at one time in a society which has no market economy or freezers. Goats fare badly in the damp conditions inland, succumbing to foot rot and eye diseases, and could have devastating effects on natural vegetation. A programme of building and stocking village fish ponds is planned but the main thrust of the Persoons' efforts is, very sensibly, concentrated on those animals which the people know; namely pigs, chickens, sea fish and, to a lesser extent, ducks.

One of the reasons for improving and introducing new protein sources in the first place was to lessen the hunting pressure on the endemic wildlife, particularly the primates. Unfortunately, the Persoons have been given permission to work only in resettlement villages from which men would rarely hunt, and so their whole approach has had to be different. Accepting these restrictions for the time being, Gerard and Jes have started to improve local chicken stock by crossing cockerels with hens imported from the mainland and then distributing the offspring. Many people, including Jane and I, have tried to keep mainland chickens in inland Siberut and all with the same fatal results. Whether the deaths were due to the inhospitable climate or to endemic diseases is uncertain and so the Persoons are introducing some new, productive genes and allowing natural selection to do the rest. They are also trying to find some simple ways of improving duck husbandry, and the efficiency of offshore fishing. More elaborate programmes of animal breeding to produce suitable strains would no doubt be possible, but neither the time nor money are available.

Gerard also decided to try bringing tame pigs to the island from Chinese-owned farms near Padang. The Siberut pigs are feral and although they will accept being enclosed for a short time, they sicken and die if penned permanently. On my third day with the Persoons the first batch of pigs arrived; two young boars, ten young sows and two pregnant sows. These were being given to the Maileppet clan, members of which had been across to Padang to see how penned pigs were housed and cared for. They had then designed and built the pens in their village using a combination of local and imported materials, and a well had been dug for a hand-pump. The pigs tumbled out of their crates and seemed to feel at home almost immediately while the villagers, all ardent pig-lovers from birth, were amazed at how they could scratch the pigs behind their ears without being bitten. One of the Persoons' assistants, an unassuming youth called Botui, even started kissing the snout of a particularly friendly young sow.

Initially the Survival International project had intended to start a vaccination programme for local pigs which were reported to suffer from occasional epidemics of disease. Gerard abandoned this, however, not only because of the impracticability of regularly vaccinating all the pigs on Siberut, but also because a veterinary surgeon sent out by Survival International found at least one of the pig 'epidemics' was in fact a nutritional deficiency. A small amount of zinc sulphide made the pigs healthy again, and people are starting to come to Maileppet for this very cheap chemical. It is so often the case that significant help need not be expensive.

Before the boat that had transported the pigs to Maileppet could return, the Muslim crew required that it should be cleansed. To do this they sought the services of a *dukun* from Muarasiberut. A *dukun* is the West Sumatra equivalent of a *sikerei*, for he has the power and knowledge to heal and also the ability to act as a spirit medium. Two of the Persoons' chickens were taken for the ceremony, one of them particularly highly prized because of its black feathers and black legs, and while these were being sacrificed on the boat to rid it of the evil pig spirits, the Persoons and I couldn't help but be saddened by the irony. The sailors and *dukun* would defend what they were doing as absolutely necessary, but defend equally vehemently their convictions, learned at school, that the beliefs of the Siberut people were primitive and had to be changed.

The next day, Gerard told me to my amazement that he had found that West Sumatra actually has to *import* sago flour from West Malaysia for the manufacture of animal foodstuffs, while the incredibly productive sago swamps on Siberut are being underexploited and their use is positively discouraged by government authorities. One of his assistants is developing a portable, mechanical sago-pith grater and Gerard is establishing outlets for the flour to enable Siberut to contribute directly to the provincial economy.

While walking around Maileppet I met some of the inhabitants of this 'new' village. Although other resettlement villages house people who have been moved, the Maileppet clan lived at this site for generations before a resettlement village was even conceived, let alone built. This makes it all the more strange that money should have been spent on building a new village when the original village is so near to Muarasiberut. But then this clan bears the brunt of many development plans because it is accessible from Muarasiberut, yet far enough away to make the officials going there feel that they are really trying. It was, for instance, the women of Maileppet who were gathered around a Minangkabau lady sent to teach them how to cook sago by boiling it with water – I dare not comment.

The fact that Maileppet was no 'new' village was abundantly clear, for nestling beneath the coconut, durian and other fruit trees were beautifully constructed traditional *umas* and large *sapos*, most built without resort to nails, that were now closed and rotting. Scarcely an energetic spitting distance from these were the serried rows of stark new houses with ill-fitting and splitting planks and corrugated iron roofs in the full glare of the equatorial sun. The West Sumatran builders managed to space these huts regularly by felling mature obstructing durian trees; a wasteful act which none of the residents would ever have contemplated. Durian fruit is eaten both ripe and unripe, forming the main meal of the day during the fruiting season, and the trees are an important element in people's wealth, often being included in a bride price. After the durians were felled one of the village *sikereis*, a cripple named Jusuf, pulled himself onto each of the durian stumps in turn and sang placatory songs to the spirits, for they are easily angered by the unnecessary destruction of useful objects. Had he not, it was conceivable that Jusuf and the rest of the Maileppet clan might have incurred the spirits' wrath and sickened.

I went to see Jusuf as he sat smoking on the veranda of the large *sapo* which he still used, in defiance of government orders. The view he had out over the sea was tranquil and beautiful; a shady grove of coconut trees with plump, brown chickens scraping busily in the grass and scattered thorny durian skins, a little tidal stream alive with air-breathing mudskipper fish and the curving white sandy beach. Beyond was the calm, mangrove-fringed bay – a mentor, a solace, a certainty in this time of change. Jusuf's wife brought out some sago, coconut and steaming boiled sea fish from the back of the dark house and I was invited to join them. Jusuf and I shared one of the wooden platters, and I was grateful to slake my thirst with some of the fishy soup from a communal spoon. The eldest daughter picked little pieces of fish off the bones and alternately suckled and fed her baby son.

When we had finished, Jusuf lit up one of my cigarettes and leant against the sides of the veranda. '*Oto, kipa,*' he said, 'what can a man like me do?' He looked out to the sea. 'I have a wife and we have five children; that is seven in all. *Uma nene, iali abeunia*, this *uma* here is big enough for all of us, and it is cool. Seven people cannot fit into that new house they have given us. The metal roof makes it very hot inside during the day because there is no shade; when it rains, *teelay*, you cannot hear yourself speak or sleep, *tak isese.*'

If people are moved closer and closer to the Siberut coast and subjected to curfews, then their disturbance to the forest must decrease and the *beelow* and other wildlife must benefit. Even if areas are logged selectively, many animals could conceivably adapt if they are given time and granted freedom from hunting. If all this is so, then surely the World Wildlife Fund and other bodies concerned with nature conservation should be glad?

No, for although ease of communication and administration are major reasons behind the resettlement programme, they may not be the only ones and, whatever the basic rationale was behind it, conservation probably didn't enter the discussion. Consider – Java is an island comprising seven per cent of Indonesia's land area but supports sixty per cent of its people who live at one hundred and fifty times the density of people on Siberut.

Consider – the total population of Siberut is less than one per cent of that of Bali, an island of very similar size and Asia's most popular tourist resort.

Consider – since the fifties a programme of internal migration known as 'transmigration' has been in operation that moves over-crowded Javanese farmers to unpopulated regions of Sumatra and Indonesian Borneo. These volunteers are provided with land, a house and some financial and agricultural assistance. Their great skill in growing rice at transmigration sites in West Sumatra has certainly shown up the local farmers and they can greatly increase the productivity of an area.

Consider – although Indonesia was self-sufficient in rice until Independence, the population growth and the failed promise of the Green Revolution have left Indonesia the world's major *im*porter of rice. She is dependent on the vagaries of the international markets and foreign relations for the provision of a proportion of the country's major staple.

So what more logical plan could there be (on paper at least) than to clear most of the remaining forest and transmigrate Javanese farmers to the interior of Siberut in order to produce the rice that Siberut people have shown themselves unable or unwilling to grow? Rice is bound to fare badly on Siberut at the moment because of the extremely high rainfall, the very acid soils and the lack of know-how and experience. But the best way to reduce rainfall (and incidentally, to reduce the fertility of the sago swamps) is to cut down the forest, for it is a vital link in the water cycle, and acidity can always be countered by occasional dressings of cheap lime.

I can see little in the insidious course of Siberut's recent history, or in the present policy affecting it, that contradicts this prognosis even if it is not the conscious intention of any one official. If present attitudes and pressures persist, I find it quite conceivable that the first transmigrants might arrive on Siberut in the next decade or so. The fact that Gerard and Art were allowed to work does not necessarily contradict this hypothesis. Both projects bring valuable foreign exchange into the country and they help to bolster the local economy, but even if they do last for five or ten years they are only passing phases when viewed against past and future history.*

The Indonesian motto '*Bhinneka Tunggal Ika*' or 'Unity in Diversity' is undoubtedly a thoroughly laudable national goal for the World's fifth most populous country, with three hundred ethnic groups using 250 languages scattered over thousands of islands spanning an eighth of the Earth's circumference. Considering what runs against it Indonesia seems an amazingly unified country and many of the people on Siberut, even *sikereis* with full body tattoos and chipped teeth, will stand up proudly and declare themselves Indonesian citizens, able to vote freely in elections. The unity is not lacking but one has to look increasingly hard for the diversity. Yet it is diversity which is the cornerstone of any stable system, man-made or natural, since it provides fail-safe mechanisms, contingencies, buffers and checks, and an array of interdependent elements all building towards their common goals of survival, growth and development.

The solution to Siberut's problems lies in education. Education outside Indonesia to demonstrate that most of the problems of the lesser developed nations are a result, at least in part, of the greed and wastefulness inherent in Western lifestyles. Education within Indonesia at the highest and lowest levels to show that wise conservation is inseparable from wise development, that conservation of natural resources is a necessity rather than an optional frippery worthy of only moderate lip-service and high-flown phrases, and that Siberut is an island to be proud of. And education on Siberut itself to generate enthusiasm for the very positive contributions the indigenous people can make to both the unity and diversity of Indonesia. If all this can be successfully put into operation, then Siberut's fascinating wildlife and people have a chance.

* October 1981. I recently received news that the Survival International project is not being allowed to continue into 1982.

Appendix

Scientific names of Siberut animals and plants mentioned in the text

Animals

Beelow	*Hylobates klossii*
Bokkoi	*Macaca pagensis*
Giant Rat	*Rattus sabanus siporanus*
Greater Thick-knee	*Esacus magnirostris*
Hornbill	*Anthracoceros coronatus*
Jirit	*Sundasciurus lowii fraterculus*
Joja	*Presbytis potenziani*
Kaloloklok	*Aeromys* sp.
Katokali	*Halictus* sp.
Loga	*Callosciurus mentawi*
Luitluit	*Calyptomena viridis siberu*
Pangolin	*Manis javanica*
Peeow	*Petinomys lugens*
Red Forest Rat	*Maxomys pagensis*
Short Python	*Python curtus*
Sikaobbuk	*Myotis monticola*
Sikaoinan	*Crocodylus porosus*
Simakobu	*Simias concolor*
Sipupaipaijutoh	*Chiropodomys karlkoopmani*
Soksak	*Lariscus obscurus*
Tamara	*Rhyncoporus ferrugineus*
Teilet n'oinan	*Rana macrodon*
Toulu	*Geomyla spinosa*
Towek	*Kuphus* sp.
Tuktuk'ake	*Phrynella pulchra*

Plants

Alalalatek	*Dendrocnide sinuata*
Alibagbag	*Endospermum malaccense*
Ariribuk	*Oncosperma horridum*
Elagat	*Dipterocarpus* sp.
Gite	*Alstonia pneumatophora*
Ipoh	*Antiaris toxicaria*
Kasuka	*Myrmecodia tuberosa*
Kataka	*Shorea* sp.
Koka	*Dipterocarpus* sp.
Latso	*Alangium ridleyi*
Payleggut	*Dillenia excelsa*
Pola	*Arenga obtusifolia*
Popokpok	*Baccaurea sumatrana*
Sigeupgepi	*Diospyros beccarii*
Tumu	*Campnosperma auriculatum*

Index

Numbers in italics indicate a drawing